THE PYRAMIDS' MYSTERIES RESOLVED

Authorized Edition
By
CLASSICAL MUSIC FOR CHILDREN
FOUNDATION

THE PYRAMIDS'

MYSTERIES RESOLVED

SCIENTIFIC SOLUTIONS TO CHALLENGES REGARDING THE EARTH'S MAGNETIC FIELD

&

CLIMATE CHANGE

1St Revised Edition with Illustrations

By

Bernard Christian Magnongui

THE PYRAMIDS'

MYSTERIES RESOLVED

SCIENTIFIC SOLUTIONS TO CHALLENGES REGARDING THE EARTH'S MAGNETIC FIELD & CLIMATE CHANGE

Printed in the U.S.A.

Written by Bernard Christian Magnongui

Second Edition Edited by Ashara Love and Jim Martyka

Cover design and illustrations by Jesse Thompson

Cover photo by sCky Art Photo & Video

First printing: December 2011 U.S.A

Second printing with illustrations: February 2013

With an update on page 241 (August 32, 2015)

First published by Classical Music for Children Foundation.

We offer a discount for educational or non-profit organizations that purchase large quantities of this book or anyone willing to donate quantities of this book to those organizations. Both paper and electronic versions of this book are available at Amazon.com and Barnes & Noble, as well as on our website:

www.classicalmusicforchildren.org

ISBN-13: 978-1492812166 (pbk.)

ISBN-10: 1492812161

Classical Music for Children Foundation (U.S.A)

To the World's Scientists, Political Leaders

And Humanity as a Whole

ACKNOWLEDGMENTS

My thanks and gratitude first go to the United States as a country and its people, for creating and maintaining the conditions and values of democracy which have allowed me to find peace of mind, and the freely available information in libraries to support the basis for these revelations which have been given to me.

My gratitude also extends to; the people of Egypt, Mexico, Peru, Equator, Bosnia and elsewhere who are the keepers of the pyramids; the scientists who have done tremendous work in trying to reveal their scientific purpose, as well as those who are looking for solutions to the reduction of our earth's magnetism; the next generation of scientists who believe and are looking for natural solutions to our world's lack of energy in order to protect our planet.

Thanks also to *National Geographic*, NASA, NOVA, Discovery Company, BBC News Science & Environment, Gillian Turner, The Register, Ehow.com, Professor Vroom Jack Rochford, Franz Lohner, John Meurig Thomas, FRS, Rudolf Gantenbrink, Frank Dorland, Technology & Science Space, Snohomish and Seattle Library staff, Christopher Dunn, David Linden, Rick Groleau, Peter Barlow, Nola Taylor Redd, Professor Mike Hulme, The Battery Council International, The U.S Department of Health, Education and Welfare, MSNBC.com staff and news service reports.

I say a very big thank you to Governor Patrick Duval of Massachusetts and to Mrs. Roslyn M. Brock, Chairperson of The National Association for the Advancement of Colored People (NAACP), for their encouragement and support in sending me a thank you and encouraging letter, upon receiving the gift of the first version of this book (2012). Also, thanks to all the scientists, researchers and news organizations I have quoted in this book.

Thank you to friends who supported me in researching and producing the first edition of this project: Sandy DuVall, Nicholas Sveslosky, Howard & Jeanette Foutchee, Gary Hanada, and Giovani Rivera.-

To special souls who have always been there for me in moments of joy and pain with my assignments with life:–Marshall Braaten, Isla Sitou, Lisa Cole, Paula Rock Madzo, Anita Phillippi, the Robert's family (Dianne Robert in San Antonio, Texas), Ken Lilly's family, Sherese Jenina Thompson, Sarah Jessica Thompson, Jesse Irving Thompson, Jesse Thompson, Gary Hanada, Luzia Lammon (mother Fatima); my sisters, brothers, cousins, nephews and uncles; to my father, Emmanuel Dominique Magnongui, and my stepmother, Pelagie Magnongui, for having provided me with a wonderful education.

I am more than indebted to my mother, M'Boumba Therese, for her unconditional love and support in my life while I am here in France. And to the people of all faiths, all nations, who come together as one to constantly explore the truth about the universe.

I would also like to say that I appreciate the help of the first editor, Ashara Love, in revising the text and adding to the clarity of the book in this second edition. She has offered her services as a labor of love, and I appreciate her dedication as a sister and friend. She has always had a deep commitment and understanding of the necessity to ask questions about our world, and has a reverence for life and spirituality that is a perfect match for this project.

I am now anxious to honor the God consciousness, the Mahanta, the Living ECK Master (Sri Harold Klemp), my spiritual Inner and outer teacher, as well as to the Vairagi ECK Masters, for their spiritual teachings, guidance and divine love, which have trained me and continuously give me the state of consciousness of learning to be in service to the God Force through divine love. The blessings I have received throughout all these revelations cannot be measured.

TABLE OF CONTENTS

"Galileo has been called the 'father of modern observational astronomy,' the 'father of modern physics,' the 'father of science' and 'the Father of Modern Science.' The motion of uniformly accelerated objects, known as "kinematics" when Galileo studied it, is now taught in nearly all high school and introductory college physics courses. His contributions to observational astronomy include the telescopic confirmation of the phases of Venus, the discovery of the four largest satellites of Jupiter, named the Galilean moons in his honor, and the observation and analysis of sunspots. Galileo also worked in applied science and technology, improving compass design.

Galileo's championing of Copernicanism was controversial in his lifetime. The geocentric view had been dominant since the time of Aristotle, and the controversy engendered by Galileo's advocacy of heliocentrism resulted in the Catholic Church's censorship of his writing. The Church maintained that Copernicus' theories were not empirically proven and that they were contrary to the literal meaning of Scripture. Galileo was eventually forced to recant his beliefs and spent the last years of his life under house arrest on orders of the Roman Inquisition." [1]

INTRODUCTION

Galileo's defense of the Copernican system, which claimed that the Earth and other planets revolved around the sun, took decades to be acknowledged and validated by even the most highly educated Europeans. Because of this discovery, humans now have a better understanding of the Universe. Today, this crucial understanding allows us to better forecast agricultural production, providing a reliable method for predicting the weather with relative accuracy.

Even though we can predict changing temperatures and storms systems courtesy of Mother Nature, our interactions with our planetary weather systems are still quite primitive. We are still unable to protect ourselves against the destructive might of Mother Nature: the excessive and unpredictable flooding, the earthquakes, the volcanoes, the tsunamis, droughts, and the tedious, long winters (snow and cold), the excessive heat which has dragged aged people to their deaths, etc.

Still remaining among the many mysteries surrounding us, is the root cause of the world's historical climate change. The only theory being seriously discussed by scientists is "global warming."

Not being a scientist but rather a student of the "Spiritual Teachings," I wish to transmit a crucial message to the scientific community and to economic policy-makers: a sort of divine "gift" concerning a theme that comes from my heart: Is there a scientific basis in the ancient construction of the pyramids? New tracks and scientific theories on this topic show that the effects of the present climate change and changes still to come, are bound to the earth's magnetic field and could be controlled by the rigorous use of our knowledge of the structures we know as the pyramids.

I had access to information inspired by my dream guidance, by my frequent contacts with the Divine Conscience known on Earth as the "Mahanta," a name that will often come back as the Inner Master or the Dream Master.

I know it may seem strange to you, but if you will bear with me, I will provide you with both inspiration and scientific facts that support my theories. I hope you are willing to maintain an open mind while I share my story and my rigorous investigation.

THE MANY SOURCES OF INSPIRATION

Despite advanced technological achievements in the 21st century, we have not yet fully understood, mastered or resolved the hidden phenomena of the Earth's core and the magnetic field that protects our planet. According to scientists, excessive radiation distributed electrically by the galaxy and solar winds has been shown to cause various cancers in the human body and other disruptions of life.

I am not a scientist or even a student of science. I am just an ordinary soul who strives daily to live his life by following the laws of life (or Spirit). I have learned to listen, as best I can, to my dreams and visions, as well as to manifest my positive dreams with the help and guidance of the God Force and my own inner guidance (the Dream Master). And more precisely, thanks to my spiritual effort, I learned to cultivate my innate capacity and to follow my interior positive guidance as well as I can.

My interest and motivation for studying the scientific purpose of the pyramids are based entirely on dreams and visions that I have had.

I was a computer technician with a certificate in Microsoft Certified Solutions Expert (MCSE), validated in 2000, from the Xincon Technology School of New York; I could not work in my profession at that time due to immigration difficulties, but I have had multiple experiences and professions that enabled me to survive in the United States since that time. I have had experience as a broker, after obtaining a license as a loan officer in the city of Houston, Texas, as a resident in the U.S.A.

Following the world financial crisis of 2008, I became unemployed. To rebound, I worked as a computer helper at Dell.

Then came a big shift in my life: a vision in a dream invited me to find work servicing clients at The Austin State School, in Austin, Texas, "a residential and training facility for adults and teenagers with developmental disabilities." Thanks to this work, and without any prior training, because of the joy I felt in bringing happiness to my patients, I developed and nurtured an ability to research and understand the beneficial effects of music on my patients.

As you can see, my motivation to clarify the scientific principles that underlie the construction of the pyramids is based on the dreams and visions that are frequently revealed to me;

curiosity pushes me to deepen their meaning.

If you don't know the world of dreams, it is understandable that you may be skeptical of my "hypotheses and revelations"; perhaps you wish to stop reading this book. However that may be, I am not the first person, nor will I be the last to offer the tracks of the information transcribed from dreams. In fact, to appeal to intuition, like a Messenger, to clarify scientific observations -- intuition which may prove to be accurate?! Why not!

In reference to history, let's pay our respects to the curious and remarkable minds of scientists and inventors such as René Descartes (1596-1650), Blaise Pascal (1626-1662), Isaac Newton (1642-1727), Montesquieu (1689-1755), Benjamin Franklin (1706-1790), Benjamin Banneker (1731-1806), Pierre Laplace (1749-1827), Michael Faraday (1791-1867), Albert Einstein (1879-1955). Let's recognize the value of philosophy descended from ancient Greece. Let's affirm the quality of knowledge bequeathed by the Schools of the Mysteries of Osiris and Isis in ancient Egypt. It is due to this inheritance transmitted through the centuries that our civilization has progressed to the present point.

In no way do I wish to imply that I consider *myself* a genius or even an inventor. Consider me an open conduit of information. I do not know why I was selected to receive this information. It is simply my duty to pass it on to you, the citizens of the world, for it to be used by those who know how to use it. Perhaps, if God's will is done, this information will provide us all with a way to minimize the impending global climate catastrophe. It is my sincerest wish that this work finds its way to, and is read by political persons and other people of note, who can further this project.

My ultimate goal: to prevent a global disaster that would generate an ineluctable social instability, and could cause our lives to revert to conditions not known since prehistoric times.

Let us go back to our review of the foremost practitioners on dreaming for scientific advancement.

For instance, let's talk about Benjamin Franklin. We may recognize that his profoundly important idea of uniting the American colonies during his lifetime was a concept that had already been implemented by close neighbors in America, The Iroquois Confederation.

The Iroquois Confederation was also known as the Five Nations of Indigenous American tribes, composed of the Mohawk, Oneida, Onondaga, Cayuga and Seneca nations.

The Iroquois Confederation League

The dreams of their charismatic spiritual leader, Deganawida, often known as the Great Peacemaker, inspired the creation of The Iroquois Confederation. By following the dreams of this great leader, the Five Nations developed an open spirit and a deep wisdom to cooperate and collaborate.

If you choose to follow their attitude of willingness to serve life, you are likely to and probably will "learn more about yourself" and others, an important value in our society, and are likely to learn more about resolving the mysteries of life. Whether you are engaged in scientific research or not, perhaps you too have experienced dreams that pertain to a scientific discovery.

In reading this book you may begin to question current theories regarding climate change, while also discovering (along with me) a possible solution for dealing with what I have now come to recognize as the main problem -- the deterioration of the Earth's magnetic field, which is closely linked to the earth's core.

In my research, following up on the dreams I received, it appears that it is the earth's magnetic field that is responsible for the climate change that is threatening the future of the world's population.

In summary, from this book, regardless of your education, you will come to learn, among other things, how the planet Earth may once have been saved by the wisdom and the scientific knowledge of previously unknown but technologically advanced civilizations, such as Atlantis and the people of Mu.

There are indeterminate records that point to the existence of these civilizations. According to legend, they ruled the Earth thousands of years ago. Even though these great civilizations are no longer physically available for us to research, fragments of their wisdom through the pyramids persists, and we will explore this shortly.

"The world will not be destroyed by those who do evil, but by those who watch them without doing anything."

--Albert Einstein [2]

"*In our every deliberation we must consider the impact of our decisions on the next seven generations.*"

--Deganawida, peacemaker and a founder of the Iroquois League [3]

"I am of the African race, and in the color which is natural to them of the deepest dye; and it is under a sense of the most profound gratitude to the Supreme Ruler of the Universe."

— *Benjamin Banneker [4]*

"I always rejoice to hear of your being still employed in experimental researches into nature, and of the success you meet with. The rapid progress true science now makes, occasions my regretting sometimes that I was born so soon: it is impossible to imagine the height to which may be carried, in a thousand years, the power of man over matter; we may perhaps learn to deprive large masses of their gravity, and give them absolute levity for the *sake of easy transport. Agriculture may diminish its labor and double its produce; all diseases may by sure means be prevented or cured (not excepting even that of old age), and our lives lengthened at pleasure even beyond the antediluvian standard. Oh! That moral science was in as fair a way of improvement; that men would cease to be wolves to one another; and that human beings would at length learn what they now improperly call humanity!"*

— *Benjamin Franklin [5]*

"Laws, in their most general signification, are the necessary relations arising from the nature of things. In this sense all beings have their laws: the Deity His laws, the material world its laws, the intelligences superior to man their laws, the beasts their laws, man his laws."

— *Charles Louis de Secondat, Baron de Montesquieu [6]*

DREAMS & SCIENCE: RENE DESCARTES

To begin your journey with me through this book, I would like to explain some of the benefits and the power of dreams, regardless of religious faith, scientific beliefs or education.

To do so, I invite you to go back in time with me to learn something about the dreams of Rene Descartes, the seventeenth-century French philosopher, mathematician and scientist, the genius who was responsible for discovering and assembling the foundational assumptions that form the basis of our modern scientific processes.

As recalled by Professor Vroom Jack Rochford, in his book entitled "Rene Descartes: A Biography," Descartes received many, perhaps most, of his scientific leads or inspirations through dreams.

"In his first dream, or nightmare, he felt perhaps it was the work of some evil genius. Phantoms appeared before him and so terrified him that as he walked through the streets, he was forced to turn over to his left side in order to reach the place he wanted, for he felt a great weakness in his right side which kept him from leaning on it...then he noticed that the wind, which had almost upset him, had become less violent...

"Almost immediately he was visited by a second dream, which consisted solely of his hearing a piercing noise, like a clap of thunder. Frightened, he opened his eyes to see a great number of sparks all around his room. This happened to him before...

"His third dream, unlike the first two, had nothing terrifying in it and was much more complicated. This time he noticed a book on his table. Who had placed it there he did not know. Opening it, he was delighted to find that it was a dictionary he thought might prove useful. At the same time he discovered another book, as surprising as the first, for he had no idea who could have put it there. Prompted by curiosity, he opened the volume and chanced *upon the line Quod vitae…*

As he was reading, a stranger appeared and gave him some verses beginning with the words Est et non… At this time both the man and the book disappeared, and yet he did not awaken…

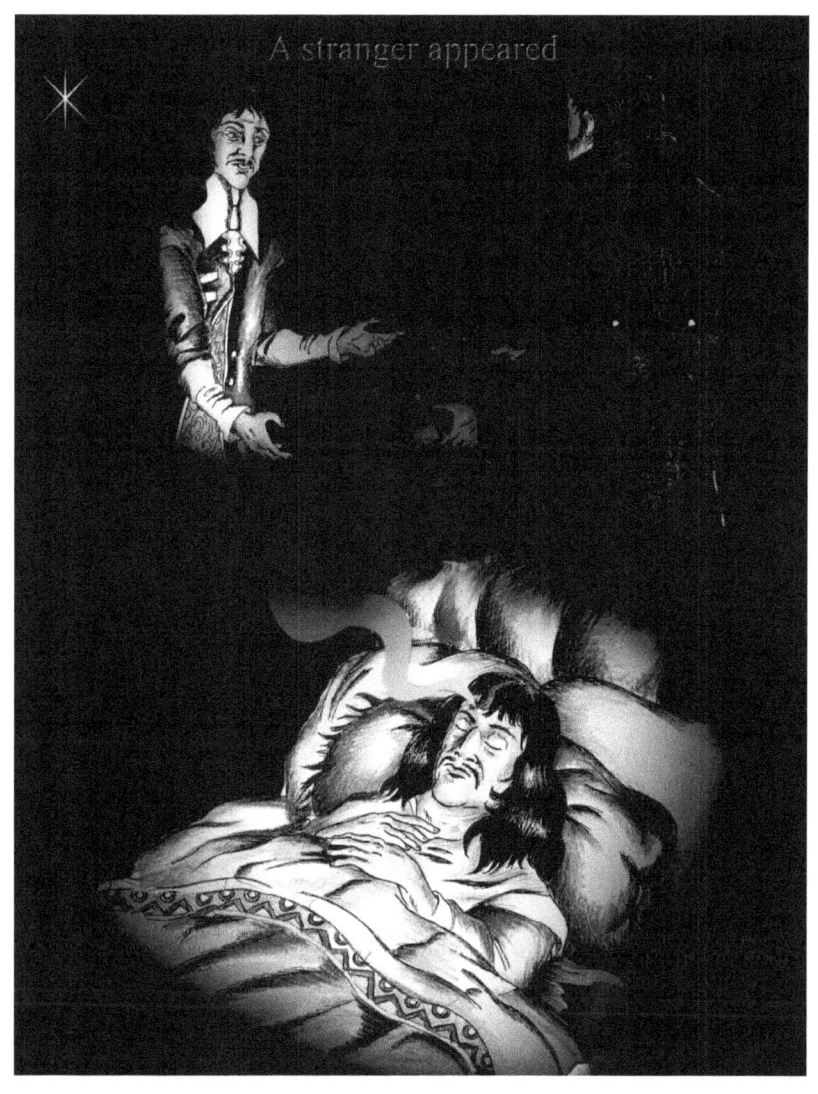

"While still dreaming, he proceeded to interpret the significance of what had just happened. He judged that the dictionary represented all of the sciences gathered together... Then, doubting whether he was dreaming or meditating, he awoke and continued the interpretation of his dream along the same lines... The poem Est et Non, the Yes and the No of Pythagoras, stood for truth and error in human knowledge. Since all the elements of his dream seemed to lend themselves to a logical interpretation, he felt that it had been sent to him by the spirit of truth which had deigned to enlighten him as to the future."[7]

We appreciate the wisdom of Descartes for having the open mind, curiosity and discernment to wisely translate his three dreams. Because of his attention and attitude toward the knowledge of dreams, as well as his search for truth, Descartes was able to accept and return the gift of the dreams bestowed upon him by interpreting them in his own words, providing him with a springboard to his world-changing scientific achievements.

This gift of the spirit became the foundational thinking upon which modern science has been able to blossom. This gift is now in our hands and homes in various forms, such as computers, which allow us to forecast the weather and other technological advances that came from inspiration.

Now let us see whether some more ancient wisdom might offer us a solution to the major problem this planet faces in the 21st Century.

"When it is not in our power to determine what is true, we ought to follow what is most probable."
— René Descartes [8]

WHO BUILT THE PYRAMIDS?

From Africa to Europe, Asia, the Americas and elsewhere, monuments still exist that cause all of humanity to wonder:

How did they come to be?

Among these structures are those that we have come to call "pyramids." Strangely, although they have captured the imagination of so many, there is much we do not know about them. For instance, who built them, what purpose they serve and why are they called pyramids?

For many centuries, either through ignorance or lack of knowledge, humans have worshiped these unexplained massive stone structures, guessing at their meaning by applying various concepts that conform to the spirituality, religious faith or fantasies of the day. Whether or not the 19th century world view explained life well to the citizens of that age, it appears that the pyramids are such an advanced technological innovation, that it is not until now that we have a full appreciation of their purpose, and the solution they represent in forestalling an inevitable global catastrophe thanks to climate change.

As I have come to understand, only few among the souls whose minds are filled with the Spirit of Truth have attempted to decode the true scientific purpose of the pyramids. But this discovery, this understanding, this attempt to explain the purpose of the pyramids to the rest of humanity, has been in vain throughout history. I am hoping you will help me spread the word about what has been explained to me.

The question that has haunted and puzzled so many, from the dawn of mankind through the 21st century is "What were the pyramids built for?"

With the help of Divine Grace, I will answer it in this book. Again, I wish to emphasize that my dreams came to me unbidden. I am simply a messenger.

Although we haven't yet discovered either artifacts or evidence proving who built them, at least two theories seem to be the most prevalent: that the pyramids were inspired and were constructed by strangers who come from faraway planets; or that they were constructed by a civilization named Atlantis that previously had colonized our planet and, according to legend, would be swallowed to the bottom of the Atlantic Ocean, and in other seas.

A HIGHER POWER AND A RULER:
Prince of Egypt, Tuthmosis IV

By reading the dream story of Rene Descartes, we learn that he, among many other scientists, had been divinely inspired. Some were inspired through intuition or imagination, others through dreams or visions or both.

Carefully reviewing the dreams of Rene Descartes, it seems reasonable to infer that, without these divine inspirations, our society would not have reached the technological advancements we consider commonplace today, with regard to the depth and breadth of scientific understanding and technology.

If it's possible that wisdom can descend to a number of humans (whether they are geniuses or ordinary people) through the power of dreams, the kind of dreams that lead to scientific breakthroughs from the fertile minds of so many individuals, these dreams may also have inspired the construction of the pyramids.

I would also like to propose that the same higher power or

God consciousness has kept the pyramids alive in our imagination and consciousness from the day of their completion until now. As I have been advised, it is now our shared responsibility to realize their scientific purpose, to re-establish their function as they may have once been used by past, unknown or misunderstood civilizations.

We must rediscover and learn their usefulness in the realm of scientific knowledge, build more of them and make use of them in order to save the Earth from a global climate catastrophe, as they apparently did in times long forgotten, before our written records allowed us to understand the exact purpose for which they were built.

Could it be that it was that same higher power of God consciousness that spoke to the Prince of Egypt Tuthmosis IV as the Sphinx? I have some questions to consider:

Why was the Sphinx built so close to the Pyramid of Giza, if not perhaps to draw our attention to a scientific reason which we were not yet ready to learn about, until now?

What if the Sphinx had been left buried underneath the desert sands of Egypt?

Would we have been able to, in our modern times, learn more about the pyramids without it?

Why was the Sphinx preserved?

To find the answers, perhaps we might attempt to travel back in time to Egypt. We may be able to learn how the Sphinx was saved from the swallowing of the sand by reading this particular story of "The Tuthmosis IV Dream Stele."

"On one of these days it happened, when the king's son Tuthmosis had arrived on his journey about the time of mid-day, and had stretched himself to rest in the shade of this great god, that sleep overtook him.

"He dreamt in his slumber at the moment when the sun was at the zenith, and it seemed to him as though this great god spoke to him with his own mouth, just as a father speaks to his son, addressing him thus: 'Behold me, look at me, thou, my son Tuthmosis. I am your father Horemkhu, Kheper, Ra, Tmu. The kingdom shall be given to you... and you shall wear the white crown and the red crown on the throne of the Earth-god Seb, the youngest (among the gods).

"The world shall be yours in its length and in its breadth, as far as the light of the eye of the lord of the universe shines. Plenty and riches shall be yours; the best from the interior of the land, and rich tributes from all nations; long years shall be granted to you as your term of life. My countenance is gracious towards you, and my heart clings to you; [I will give you] the best of all things.

"The sand of the district in which I have my existence has covered me up. Promise me that you will do what I wish in my heart; then shall I know whether you are my son, my helper. Go forward. Let me be united to you. I am...

"After this [Tuthmosis awoke, and he repeated all these speeches,] and he understood (the meaning) of the words of the

god and laid them up in his heart, speaking thus with himself..."
[9]

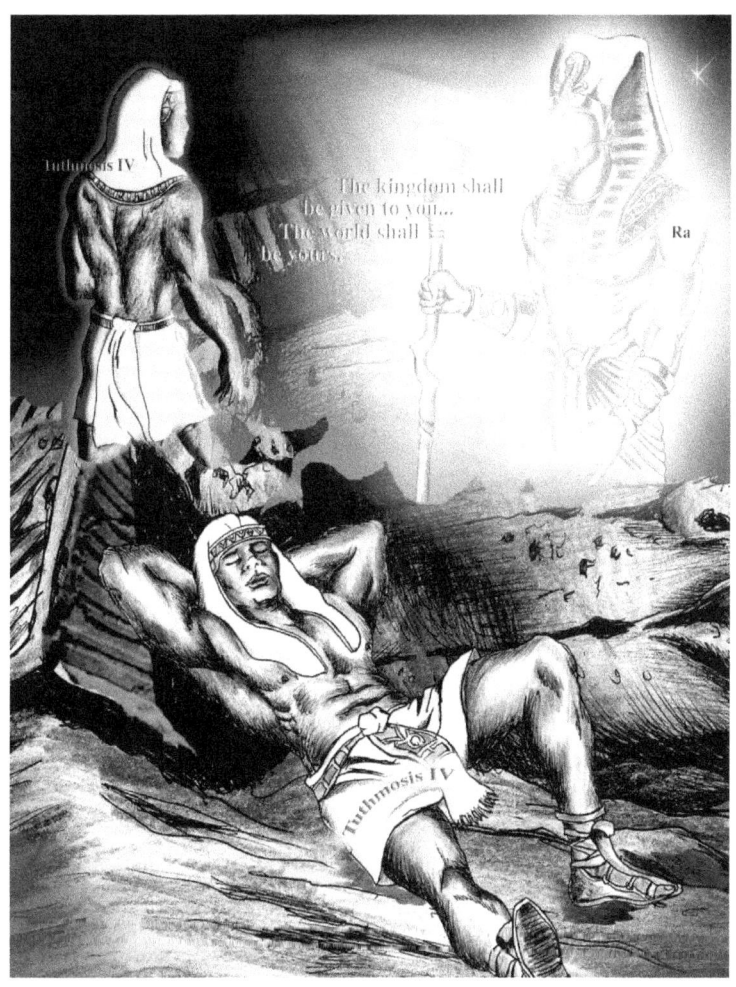

And so, the sand surrounding the Sphinx was cleared up. This protected it from being forever buried underneath the desert sand.

Without the dram of Tuthmosis, and without his wisdom in listening to and accurately translating the *positive* request of Ra in his dream, we in the modern world would perhaps never have heard of the existence of the Sphinx.

Perhaps in the near future, we will realize its scientific connection with the Pyramid of Giza.

Perhaps other dreams might have come from unknown people involved in seeking to resolve the mystery of the pyramids, and perhaps other pyramids have survived centuries of burial in desert sand. Lately we discovered the existence of other pyramids in Sudan, Mali, in Bosnia, in Mauritius, in the Canary Islands, as well as 250 in the U.S.A. (Thanks to the research of Dr Sam Osmanagich). Who can ever know what was discovered in the distant past, or what might be discovered in the near future?

But we *do* know, according to the dream account of the Prince Tuthmosis, that the invisible voice of the divine sometimes speaks to a person for the benefit of all creation, through the medium of dreams; the fundamental goal is to bring a positive change of personal or collective consciousness in view of the evolution (morals, intellectual, artistic, social, scientific, philosophical, spiritual or religious, political…etc.) of all Creation, which means all of humanity. It can be revealed to any person, without distinction of faith, life or social standing. Regardless of faith, social life, or position in society, a dream can come to anyone.

So, my dream stories actually have a historical precedent. Now, allow me to explain, dear readers, how this came about.

HOW IT ALL STARTED

One of the first dreams I could relate to the pyramids occurred at the beginning of the month of August, 2011 while I was busy compiling research I had gathered to write the upcoming sequel to my first published book entitled "The Power of Musical Sound: How Music Affects Our State of Mind, Health and Society."

If you wish to obtain a copy of the book, contact us on our web site: www.classicalmusicforchildren.org

I was busy selecting the best articles on the subject of the effects of vocal words in music (lyrics). That night, after a period of contemplation and sleep, I had a dream that made no sense to me at all.

I heard the name "Montesquieu" called out although there was no one else there, as if I were being called by another name. The dream made no sense to me...I was a little bit confused, because my friends have always called me either Bernard or Christian; and since there was no else around except me in the dream, I understood that the Spirit of Truth was perhaps drawing my attention by calling me by that name. (See picture of Montesquieu bellow) [10]

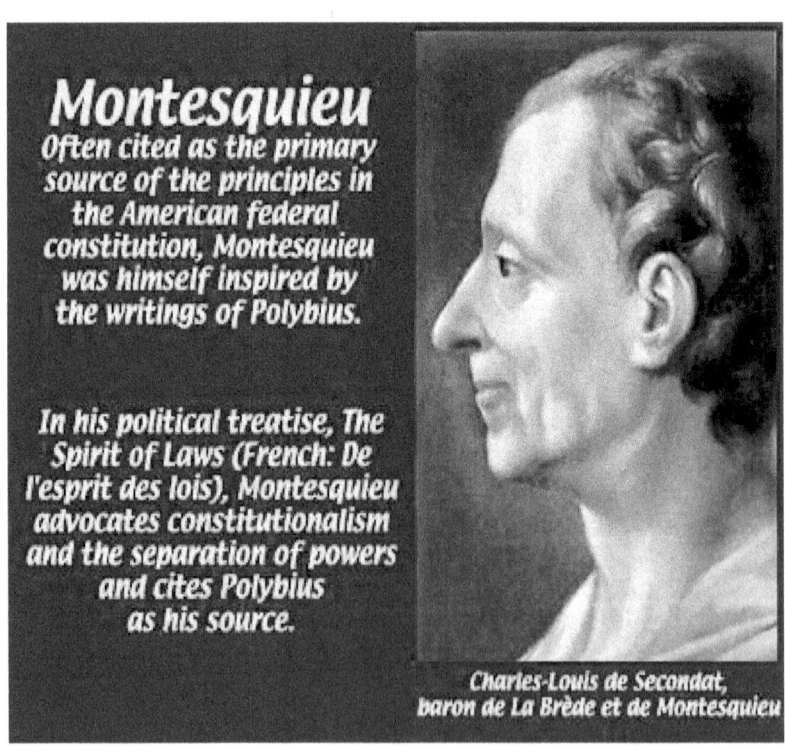

Montesquieu
Often cited as the primary source of the principles in the American federal constitution, Montesquieu was himself inspired by the writings of Polybius.

In his political treatise, The Spirit of Laws (French: De l'esprit des lois), Montesquieu advocates constitutionalism and the separation of powers and cites Polybius as his source.

Charles-Louis de Secondat, baron de La Brède et de Montesquieu

After that, in the dream, a geometrical square figure was drawn in front of me, with someone asking a question about its perimeter.

When I awoke, I wrote down the dream and tried to figure out what the God Force or my inner guidance was telling me. (In my spiritual tradition, the God consciousness is known as the Mahanta, or the Inner or Dream Master.) This dream was difficult for me to decode right away. Then I remembered that while in college, back home in Africa, I was taught about Montesquieu, a French Philosopher, well known as "Le Baron de Montesquieu," but I had forgotten what he had accomplished for the world.

I went online to search for his name and in time I found and read his famous work, "The Spirit of Laws." As I learned more about Le Baron de Montesquieu, I discovered that the leaders of the French and American Revolution were inspired by his writings.

I soon realized that the dream about Montesquieu was also guiding me to learn how to apply some of his thoughts about music education to the book I was writing about the beneficial effects of music. To be honest, I have not fully understood why I was called by that name. Maybe it is scientifically linked to the construction of the pyramids.

The second element of the dream, regarding the geometrical figure of the square and its perimeter was still a mystery to me. I relaxed and stopped thinking about it, as I knew that I would grasp its meaning when the time was right; and that it could be in days, weeks, months or even years. (See following picture.)

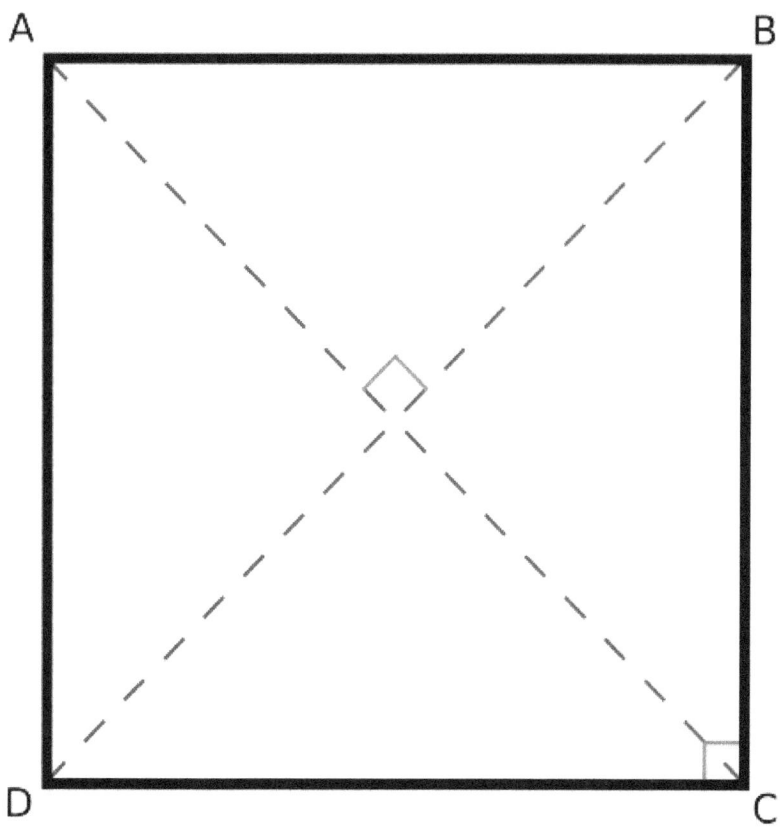

PERIMETER OF THE SQUARE = SIDE X 4

THE GIFT OF THE PYRAMID

Then on August 31, 2011, my birthday, I wondered where I should go to celebrate. It was the first time since I had come to the U.S. that I felt happy enough to celebrate in this beautiful continent, symbol of liberty and equal rights. With no money and no girlfriend at the time, I decided to let it go as usual. That night, an unexpected out-of-body experience, or perhaps we might call it a spiritual journey, occurred during my sleep.

I would like to emphasize one fact here: as a student of the spiritual teachings of Eckankar, the religion of the Light and Sound of God, I have a spiritual teacher known as the Mahanta, the Living ECK Master [11].

He has been my inner and outer teacher for approximately 25 years, in this lifetime. When he appears in my dreams to teach about God's love or the spiritual principles of life or their lessons for my spiritual growth, he is called the Mahanta, the Inner or Dream Master.

In ancient Egypt, the Mahanta as a God consciousness was known variously as God Re (Ra), the Sun God, Osiris and Isis. Gopal Das (the author of the original book of the dead, and the founder of the mystery school of Osiris and Isis), etc; in the Americas (Native Americans; Indians), he was called the Great Spirit; he was also known in the former Mexico (Mayans, etc.)

under the name of Quetzalcoatl and Kukulkan, also as Zeus by the ancients Greeks, Jupiter by the ancient Romans, and other nominations (God of science, God of wisdom, etc) in other civilizations. That night, when the Dream Master came to me in the dream state, I was surprised to see him, since he had not visited me in my dreams for quite awhile.

The Mahanta, as the God consciousness, often appears to souls regardless of their religious faith in the dream state as a blue star. But often, to his students such as me and others around the world, or even to those souls who are not members of Eckankar, he appears in the matrix of the ECK (the Holy Spirit) as the Living ECK Master. So the Mahanta consciousness is always associated with the six-branch star as shown by the picture above. But in my dream he came to me as the Mahanta, the Living ECK Master (the Dream Master or the Inner Master).

So he invited me to travel with him, by taking me out of my physical body into the soul (or Spirit or light) body; and together we found ourselves in Egypt inside the Pyramid of Giza. It was quite an unexpected choice of places to visit with the God consciousness, since I had never had any interest in the pyramids. Nor had I visited the pyramids in this lifetime. "What spiritual truth could I learn in this massive building of stone?" I wondered to myself, especially because I had never had any interest in their existence. Yet I was there, surprised by the choice of visiting place by the Divine supreme consciousness.

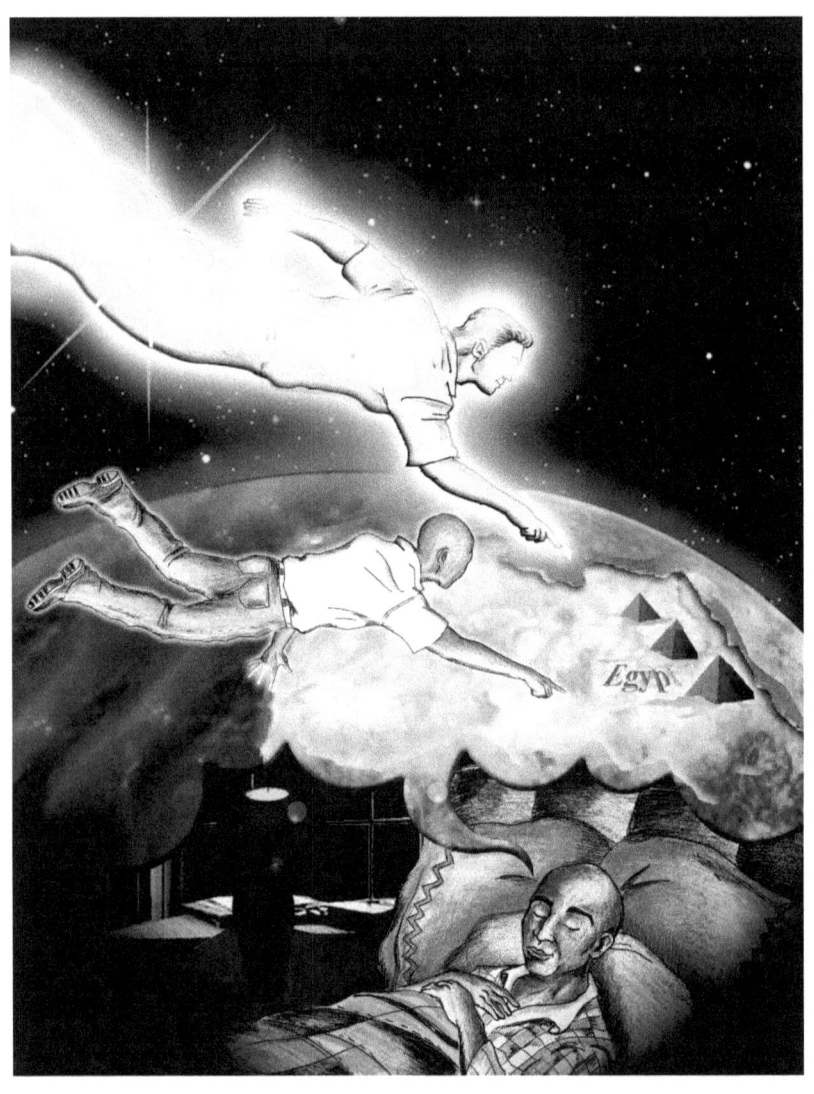

From the inside of this spiritual consciousness, I looked at the top of the Pyramid of Giza and noticed there was no roof, but instead there was an opening showing a portion of the universe with its many stars and geometrical figures that I could not understand.

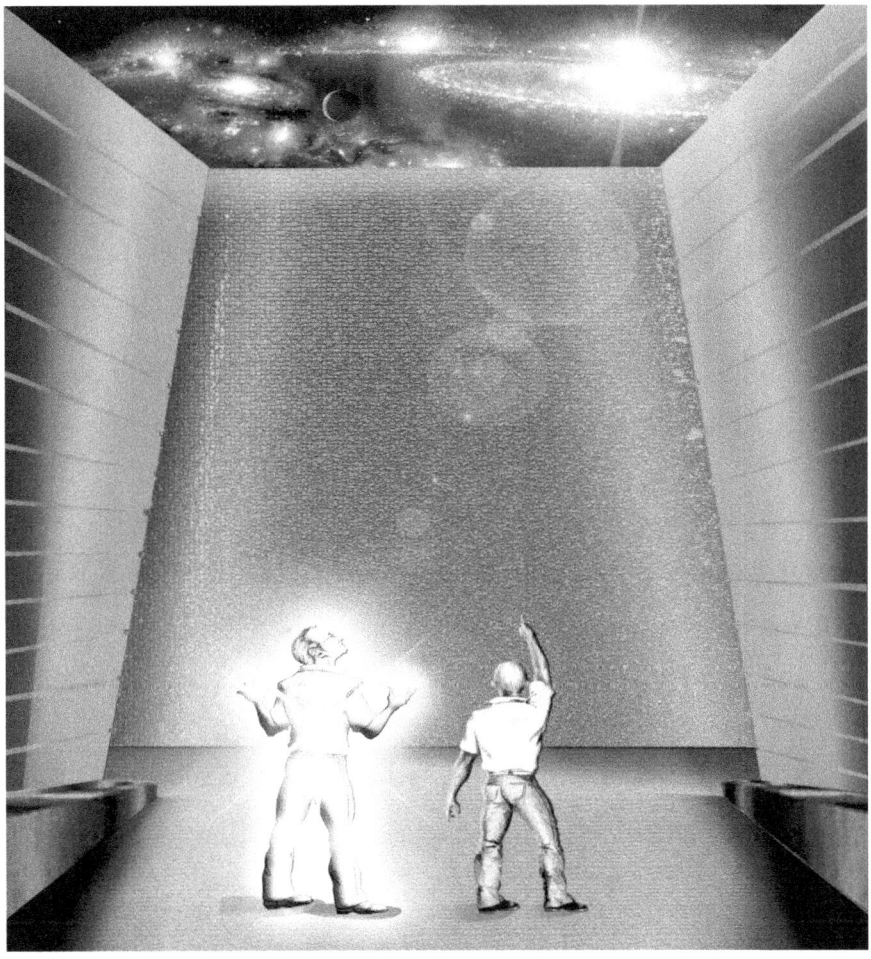

They were complex geometric figures, and not being a mathematics student, I found them more interesting and perplexing than anything I had ever seen or considered.

Inside the Giza Pyramid, the God consciousness of heaven (the Dream Master) did not say anything to me. He was just quiet and allowed me the enjoyment of the experience of being in this structure, discovering its wonders for the first time in this lifetime. I felt so happy, amazed and grateful to be invited by the Inner Master into this mysterious building of stones. Suddenly, a voice at the top of the pyramid, which I will call the Spirit of Truth (The Holy Spirit, the Force), began to speak and we both quietly listened to it. The voice was talking to me.

When the voice finished speaking, the Dream Master started speaking for the first time since we arrived. As the Master spoke about the pyramid I listened carefully without interrupting him with questions, although I wanted dearly to ask him many.

The following morning, when I woke, I didn't quite manage to believe what had happened, and I lay in a petrified state on my bed, trying to absorb and to understand the meaning of my experience. It was the most unusual place I have ever visited with the Dream Master; yet I was grateful for the visit.

Now it was time to fully understand and accept what I had experienced with my whole mind.

Wow, I thought, this is a huge, beautiful and strange gift from the Dream Master to me. A pyramid. Finally, after having absorbed the experience with my all physical senses, I understood that, while it was an enormous gift, I also knew that by the type of experience I'd had, named "out-of-body" (expansion of the consciousness of the soul), that the Conscience of the Divine was transmitting something very profound that I would have to decode by myself. In spite of all the wisdom I had received during this encounter in the pyramid, I didn't understand everything, and this was probably for a very good reason.

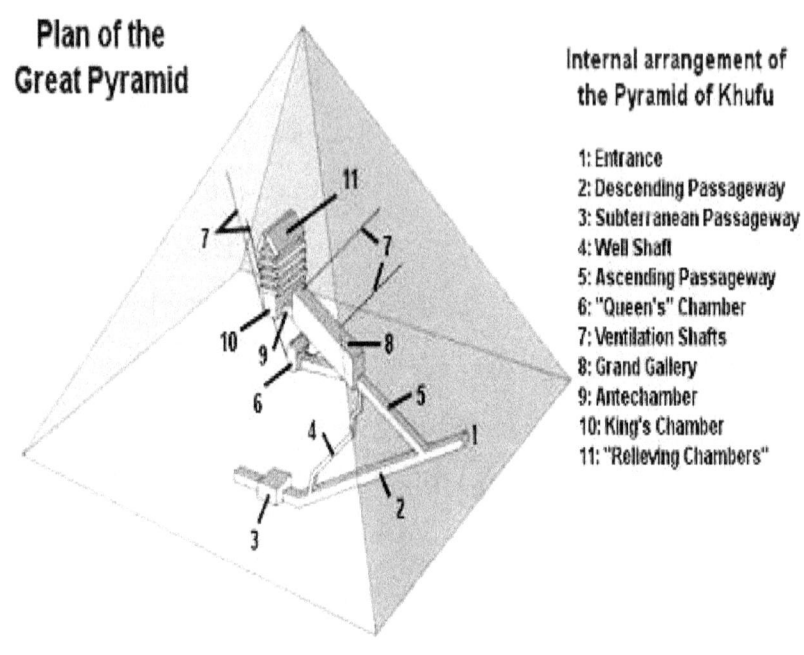

Plan of the Great Pyramid

Internal arrangement of the Pyramid of Khufu

1: Entrance
2: Descending Passageway
3: Subterranean Passageway
4: Well Shaft
5: Ascending Passageway
6: "Queen's" Chamber
7: Ventilation Shafts
8: Grand Gallery
9: Antechamber
10: King's Chamber
11: "Relieving Chambers"

Picture [12]

Consumed by a burning spiritual curiosity, especially after what I had learned from the voice and from the Inner Master's explanations about the pyramid, I decided that it was important, maybe essential, to decode the meaning of the dream.

The Inner Master had not, in fact, told me exactly what he was expecting from me. I was already writing my second book, which he had inspired me to do. I didn't think he was asking me to stop writing that book so that I could write about the pyramids. And if, in fact, he was asking me to write about the pyramids, what for?

ANSWERING THE INVITATION

After contemplating the invitation of the God consciousness inside the pyramids, I realized that I had indeed been given a new spiritual assignment. I stopped compiling the second book I was writing, to focus on learning about the pyramids. Honestly, I wasn't ready for another huge spiritual adventure, since I was now living in Washington State and I hadn't found a job after having applied to many places. But I started to realize that it was Spirit that was preventing me from finding work. I believe this was on purpose, so I could be fully available to learn about this new assignment.

Now it looked like I would be spending most of my time doing research on the Pyramid of Giza. And since it was a birthday gift from God, I decided to accept it willingly, by spiritually challenging myself to take on the assignment.

I went to the local library in Snohomish, as well as to the main library in downtown Seattle, and rented several DVDs and books related to the pyramids.

As I spent time reading and listening to previous explorers' works on the Pyramid of Giza, I became aware of something that drew on my spiritual sense: the *sun and the existence of pi*. My observational acuity as well as my spiritual sense had developed over the years of my spiritual studies. These seemed to point to the sun as the primary connection to the Pyramid. It certainly cropped up often enough as I learned about the pyramids. Perhaps I just know how to pay attention. (See following picture of the sun's composition)

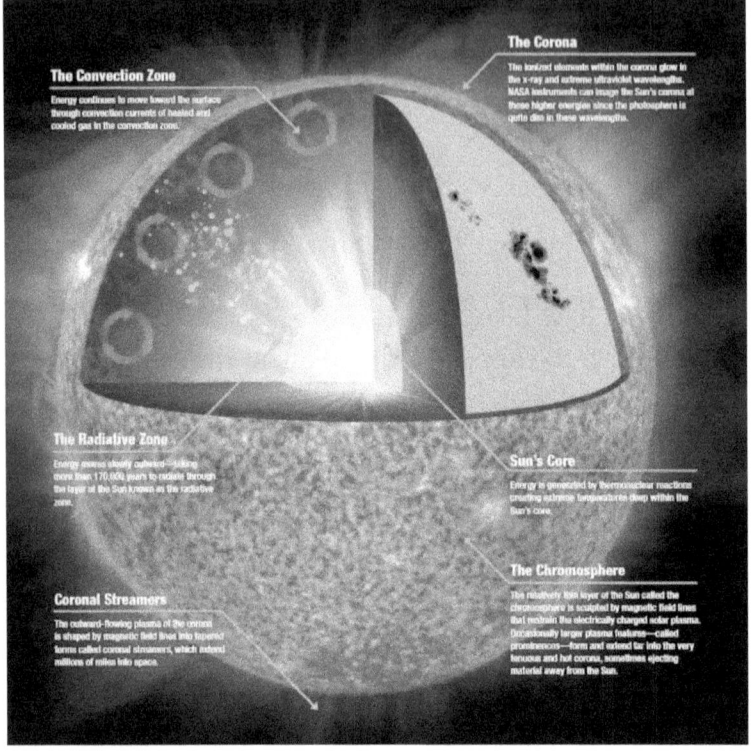

Picture [13]

"The Sun is composed of hydrogen (74% of the mass or 92,1% of the volume) and of helium (24% of the mass or 7,8% of the volume). Although the Sun is a medium-sized star, it represents only about 99,86% of the mass of the solar System. Its shape is nearly perfectly spherical, with a flattening at the poles estimated at nine micros, what means that its polar diameter is smaller than its equatorial diameter of only ten kilometers." [13A] (Translated

from French)

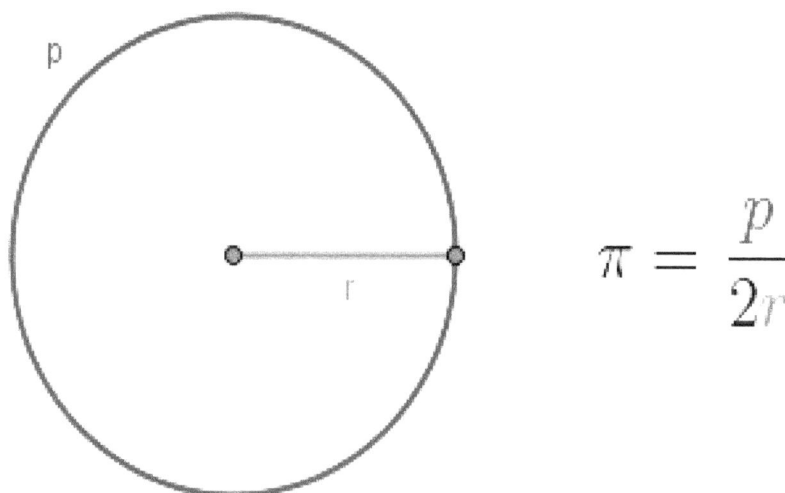

$$\pi = \frac{p}{2r}$$

"The formula giving the p perimeter of a circle in relation to its radius, is learnt in school from the earliest age: p=2 π r, where π *is* a number whose value equals "three point fourteen". By reversing the equation p_=_2 π r, one finds that, for any circle, the value of π is p/2r. The original calculations for determining π are used as a basis: if one approaches p and r sufficiently, one can find an interesting value for pi. $\Pi \approx 3.141592$" [13B] (Translated from French)

Later I learned more about the geometric square figure and its perimeter, used to build the Pyramid. It was because of this additional information about the structure of the Pyramid that I began to realize the validity and consequence of the dream I'd had. I became even more enthusiastic and eager to learn more about the Pyramid of Giza. (See the following picture 14)

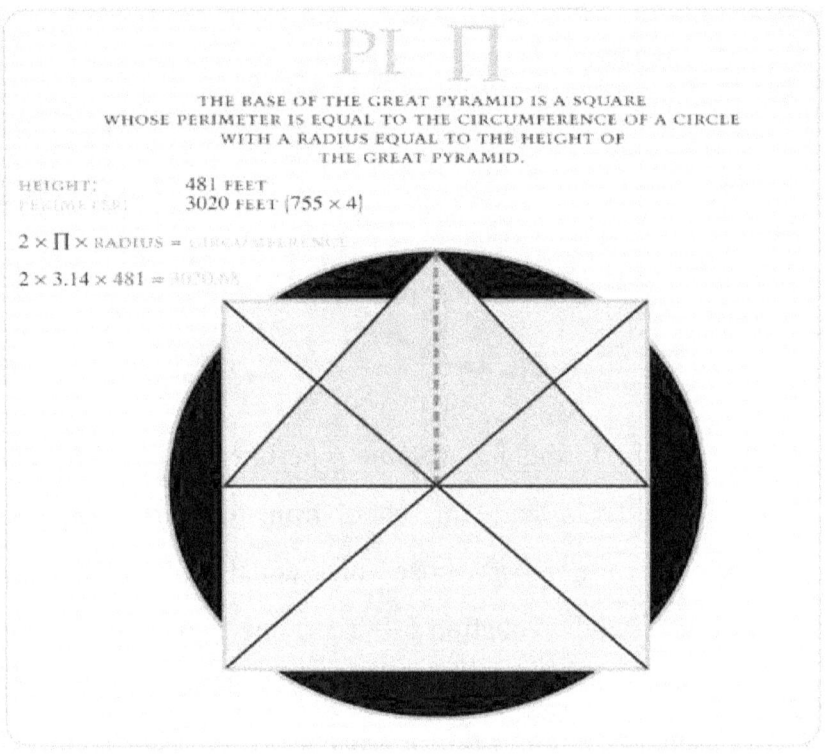

Even though I had never been to Egypt, except through the out-of-body experience with the Dream Master, I found online video materials produced by other curious seekers. I was particularly enthralled by the materials from NOVA, which greatly enhanced my education on the subject.

These educational programs on the subject of the Pyramid of Giza allowed me to view the inside of the Pyramid in streaming videos. I easily recognized the location within the Pyramid which I had been fortunate enough to visit in the Dream state with the Dream Master [11].

The elements of my dream really fell into place as I became more conscious of the nature of the gift the Dream Master had given me. The God consciousness was offering me the enormous gift of resolving the mysteries of the Pyramids. Even now, when I say this, I cannot believe it is me saying it. But I truly believe that by divine guidance, this is the case. But now I sincerely give thanks to the divine inspiration for being assigned this task. Despite all the difficulties and obstacles that I had to overcome in the United States while transmitting this message to American Scientists and politicians as well as to the media, I am still not discouraged, at the cost of having sacrificed my material life (my work, my social life, etc.) for two years. And since I could not get support in the United States, to continue to spread the message to the world, on September 07, 2013, I left the city of Seattle (Washington State) for France (Europe), and later Africa, with the goal of finding some sponsors in the political and scientific world, to help me with this project, as well as to learn more about the science of the Pyramids, with the ultimate goal of serving humanity.

Filled with gratitude for being chosen by the Higher Power from among the students of the Inner Master on Earth, I applied myself more and more diligently to learning about the Pyramids. Eventually, as I committed myself over the following weeks, the Spirit of Truth (which you know as the Holy Spirit, the God Force, the comforter, the Bani, Divine Force, etc.) revealed to me three fundamental scientific purposes of the Pyramids.

I received this information over two separate nights, and it took me almost three months to fully understand each aspect. By the end of October 2011, after having attended the yearly spiritual inspirational talk of the God consciousness on Earth, the Living ECK Master, Harold Klemp in Minneapolis, I found I was already able to understand the specific terminology of the three scientific aspects of the pyramids, as they were revealed to me in these divinely inspired dreams.

THE SCIENTIFIC CONSTRUCTION OF THE PYRAMIDS

As transmitted to me in the dream state, I was told that the pyramids were built using the concept of a battery.

Let's stop here. I am not a technical person. I sincerely hope that the people reading this book have the technical knowledge to be able to understand this discussion, but at the beginning of this process, I did not. In case you don't have any scientific background and are reading this out of your own curiosity and interest, I suggest we first break down a very basic question: What is a battery?

The best answer comes in David Linden's "Handbook of Batteries."

"A battery is a device that converts the chemical energy contained in its active materials directly into electric energy by means of an electrochemical oxidation-reduction (rod) reaction states."

"This type of reaction involves the transfer of electrons from one material to another through an electric circuit. In a non-electrochemical redo reaction, such as rushing or burning, the transfer of electrons occurs directly and only heat is involved…

"While the term 'battery' is often used, the basic electrochemical unit being referred to is the 'cell.' A battery consists of one or more of these cells, connected in series or parallel, or both, depending on the desired output voltage and capacity.

"The cell consists of three major components:

> *The anode, or negative, electrode—the reducing or fuel electrode—which gives up electrons to the external circuit and is oxidized during the electrochemical reaction.*

> *The cathode, or positive, electrode—the oxidizing electrode—which accepts electrons from the external circuit and is reduced during the electrochemical reaction.*

> *The electrolyte, or ionic conductor, which provides the medium for transferring electrons, as ions, inside the*

cell between the anode and the cathode. The electrolyte is typically a liquid, such as water or other solvent, with dissolved salts, acids, or alkalis to impart ionic conductivity. Some batteries use solid electrolytes, which are ionic conductors at the operating temperature of the cell." [15]

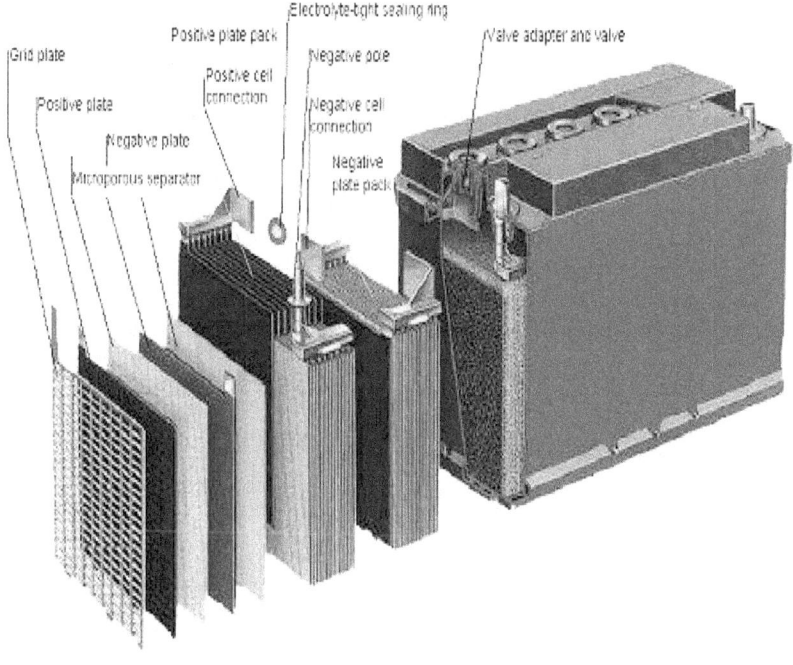

Picture [16]

From my investigation into the nature of the pyramids, I

started to recognize that the positive (+) and negative (-) poles or electrodes are represented by the pyramid's King and the Queen Chambers respectively. My investigative research also confirmed the implication in my dream that the Pyramid serves as a battery.

Christopher Dunn is the author of a best-selling book on the pyramids: "The Giza Power Plant." He is an engineer with over 45 years of experience. In the last 28 years, he has published numerous articles and has appeared in several documentaries discussing his work. He is also an advisory board member of the "Great Pyramid of Giza Research Association."

His thesis provides a sound technical basis for my dream-inspired theory. From Dunn's online article entitled, "The Evidence Leading up to Gantenbrink's 'door'," I found proof that elements of our modern day battery are also found in the Pyramid of Giza. As stated by Dunn:

"Don brings up a good point when he mentioned the chambers turning into salt as the result of interaction between hydrochloric acid and the calcium carbonate (limestone) composition of the chamber. This chamber is the only chamber that was noted to have a buildup of salt on the walls and the ceiling. It is reported to have been built up to about an inch thick

in places. In 'The Giza Power Plant' I present the results given in 1978 by the Arizona Bureau of Geology and Mineral Technology which did a chemical analysis of this salt. He found it to be a mixture of calcium carbonate (limestone), sodium chlorate (halite or salt), and calcium sulfate (gypsum, also known as Plaster of Paris. Patrick Flanagan, Ph.D, collected the samples and certified its origin.

The features found in the King's Chamber led me to propose the use of hydrochloric acid in the Queen's Chamber. The features in the Grand Gallery led me to understand the function of the King's Chamber. The features in the Queen's Chamber [25] indicate that a chemical reaction was taking place there. The hypothesis rises or falls on the evidence found in these areas."[17]

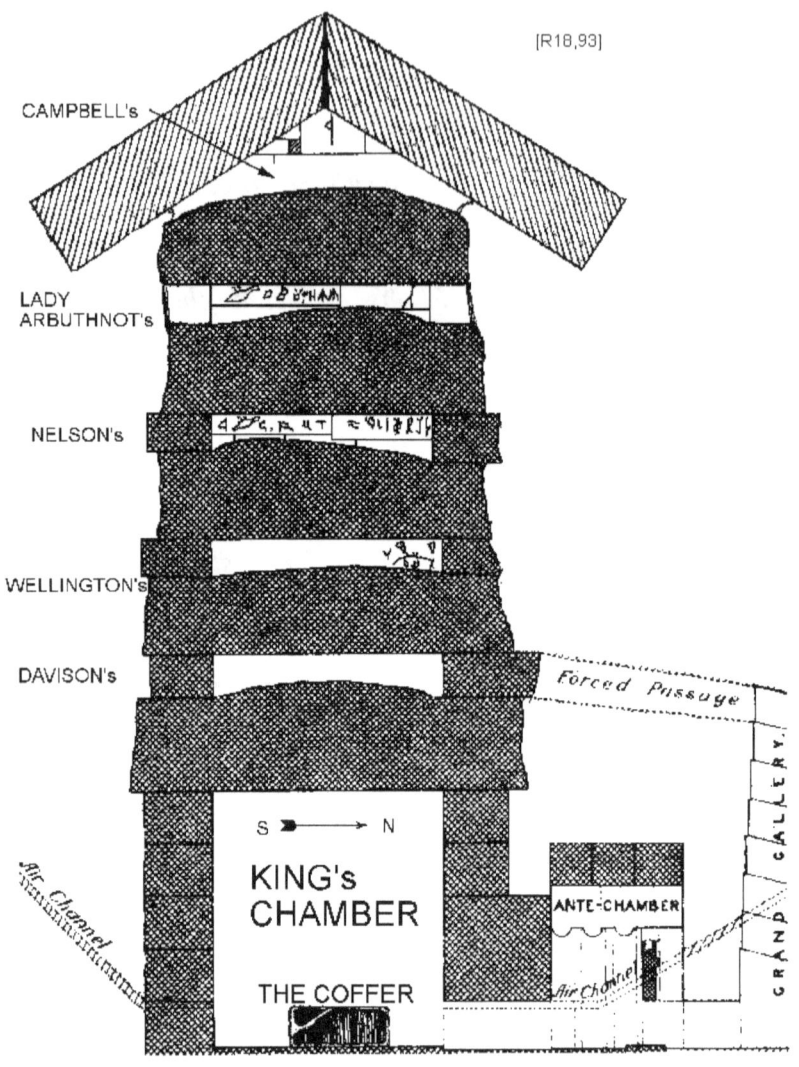

Picture [18]

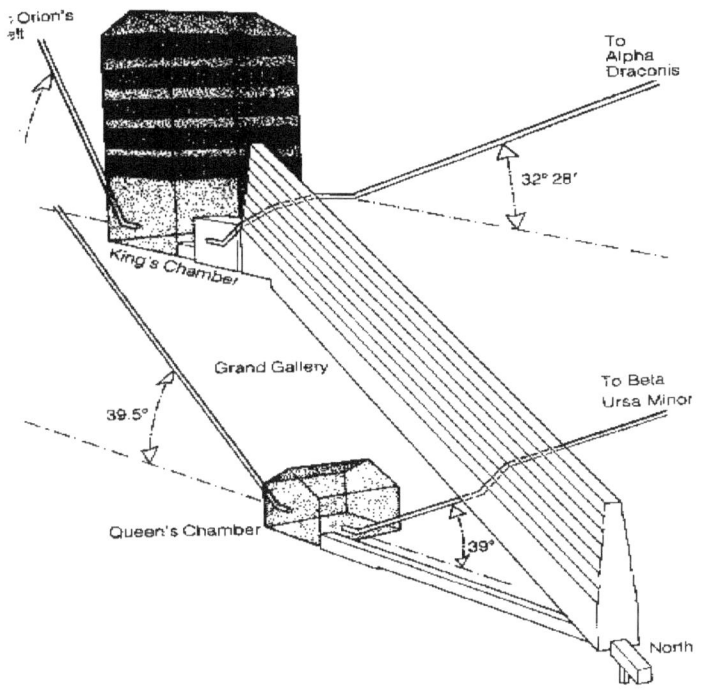

Picture [19]

To further strengthen this theory of the pyramids as batteries of different voltages and capacities, especially the Pyramid of Giza, I'll turn to another expert source on building batteries. According to the Battery Council International:

"Batteries are made of five basic components:

A durable plastic container.

Positive and negative internal plates of lead.

Plate separators made of porous synthetic material.

Electrolytes, a dilute solution of sulfuric acid and water, better known as battery acid.

Lead terminals, the connection point between the battery and whatever it powers.

"The manufacturing process begins with the production of a plastic container and cover. Most automotive containers and their covers are made of polypropylene. For a typical 12-volt car battery, the case is divided into six sections, or cells, shaped somewhat like one row in an ice-cube tray. The cover is dropped on and sealed when the battery is finished.

"The process continues with the manufacture of grids or plates from lead or an alloy of lead and other metals. A battery must have positive and negative plates to conduct a charge.

"Next, a paste mixture of lead oxide—which is powdered lead—and other materials such as sulfuric acid and water, are applied to the grids. Expander material made of powdered sulfates is added to the paste to produce negative plates.

"Inside the battery, the pasted positive and negative plates must be separated to prevent short circuits. Separators are thin sheets of porous, insulating material used as spacers between the positive and negative plates. Fine pores in the separators allow electrical current to flow between the plates while preventing short circuits.

"In the next step, a positive plate is paired with a negative plate and a separator. This unit is called an element, and there is one element per battery cell, or compartment in the container. Elements are dropped into the cells in the battery case. The cells are connected with a metal that conducts electricity. The lead terminals, or post, are then welded on.

"The battery is then filled with electrolyte—or battery acid—a mixture of sulfuric acid and water, and the cover is attached. The battery is checked for leaks.

"The final step is charging, or finishing. During this step, the battery terminals are connected to a source of electricity and the battery is charged for many hours." [20]

If we look carefully inside the Pyramid of Giza or other pyramids located elsewhere around the world, we will find evidence demonstrating the science of pyramids serving as batteries. The more interesting question is *why*. We will discuss this a bit later.

First, we must recognize that these ancient battery arrays were built using sophisticated technical knowledge of the properties of granite and other stones, which contemporary society has not fully realized or discovered, but is just today learning. Let us look closely at the construction of the pyramids.

With what appears to be advanced technological knowledge of geology, geosciences, chemistry, civil engineering, geophysics, alchemy, geomagnetism, physics and astronomy, etc, the builders of the pyramids used a combination of various volcanic rocks, which are those rocks formed from magma from an active volcano, as well as quartzite stones, such as granite (selecting different colors apparently because of their different chemical properties), limestone and other stones.

This combination of igneous, clay, granite and limestone stones tells us about the builders' level of knowledge of the Earth's inner core: its formation and the chemical composition of its mineral components.

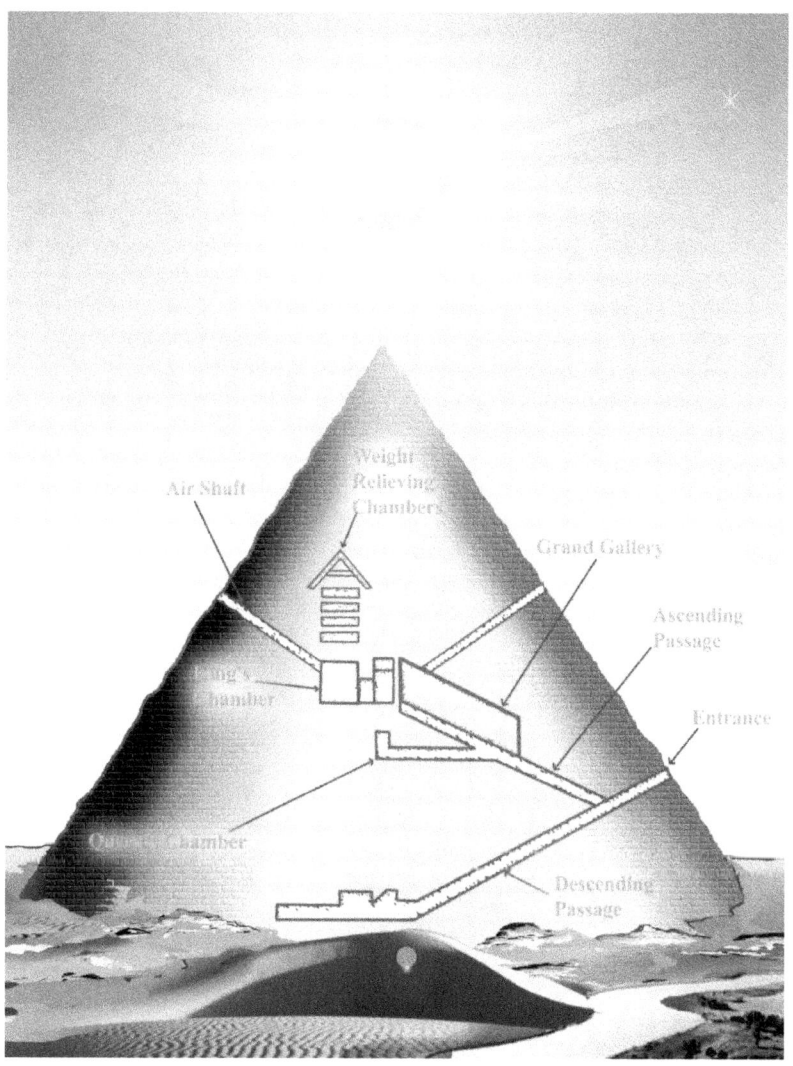

THE COMPOSITION OF THE EARTH'S CORE

According to Vincent Deparis, a researcher at the home of 'Sciences of Man of Grenoble (France)', in one of his articles titled: *"Histoire d'un Mystère: l'Intérieur de la Planète-Terre* (History of a Mystery: the Inner of Planet Earth), tells us that René Descartes "*is in 1644, the first person to imagine the underground world. For him, the Earth is a former Sun that underwent a peculiar evolution. At the center, one finds a core of solar matter, covered by a compact layer of the same matter as the solar spots. Then comes a dense layer of earth, a layer of water, a layer of air and a new lighter layer of earth that hovers above the emptiness like an arch.* "

He continues: "*The Earth of Descartes is therefore hollow! The external layer is however in unsteady balance. Dried by the Sun, it crackles, and ends by collapsing in an unequal manner onto the inner planes, expelling the water that forms the oceans. Descartes describes the genesis of the Earth and its internal structure thus. He tells us how the mountains are formed, by downfall, at the time of an immense original global disaster.*

"[21A]

Also, it obliges us once again to ask the disturbing question: "Who therefore had so inspired René Descartes that he could imagine the composition of the earth's core? From where did this knowledge come?" In our era, Descartes was the one who introduced the concept of the interior of the earth. According to historic writings, Socrates incited his students to study the geology of the earth; did he make allusions to the inside of the earth? If yes, where would have he learned this knowledge, if not by his multiple journeys to Egypt of the Pharaohs; did he study at the spiritual schools of Osiris and Isis?

The irrefutable proof that Socrates studied with the Egyptians priests is found in a quote that people ascribe to him: "*Know thyself.*" This same quote is found in the esoteric teachings of the Egyptians priests, whose spiritual lessons were based on the concept of the immortality of the soul; from which came the concept "know thyself" not as a human, but as a divinity living in a human body. In addition, the history of Ancient Egypt reveals that this quote was engraved on the walls of the temples of Egypt.

Could it be coincidental that René Descartes and the Greek Socrates and the Egyptians priests all had the same inspiration of the knowledge of the Earth's interior if it was not bound to the survival of their nation and of the planet?

Talking about the Earth's core.

A question was asked to Dr. Ken Rubin, Assistant Professor at the Department of Geology and Geophysics, at the University of Hawaii on the subject of the Composition of the Earth's Core. (Refer to picture 21). He was asked, how do scientists know what is in the core of the Earth? The professor gave the following reply:

"Well, we have a pretty good idea from a variety of indirect measurements and reasonings:

"First, we know the overall density and mass of the Earth based on measurements of how the Earth perturbs the orbits of other planets and the moon.

"Second, we know the overall density of the various layers of the Earth based upon the way in which seismic pressure waves (compressional waves created by earthquakes) move through the earth to arrive at locations remote from the earthquake source.

"Third, by examining a second type of seismic wave (a shear

wave, that is equivalent in motion to a back and forth rubbing of one's hands together) we know that the outer part of the core is liquid, even though it is at immense pressure from being underneath so much rock. Shear waves can't travel through liquids.

"Fourth, we know the overall composition of the Earth by examining the bulk chemical composition of the Sun (by examining its light spectrum) and by analyzing a class of meteorites known as Chondrites (which have similar composition to the Sun and are believed to be similar to the material from which the Earth accreted).

"Fifth, we know the composition of the Earth's crust and its mantle, by examining samples of them. For the lower mantle, we use experiments of the effect of pressure on upper (shallow) mantle minerals to predict the mineralogy of the lower reaches of the mantle. We then pass seismic waves through it in the lab to see if our experimental rocks match the observations.

"Six, now that we know the size, mass and composition of the whole Earth, its crust, and its mantle, we can construct a balance sheet of materials and see which chemical elements aren't in the crust (including atmosphere and hydrosphere) or mantle that we know should be on the Earth. These must be in the core.

"Seven, to aid us in our assessment, we recall that we need metallic elements in high concentration somewhere in the interior of the Earth to generate our magnetic field. Also, this metal must be able to be in the liquid state even at very high pressures.

"Adding all this up, we find the core is predominantly Iron metal (Fe). We find it has a significant amount of the element Nickel (Ni, about 4%) and a light element to make it less dense (about 10% by mass). This light element is either mostly oxygen or sulfur, with the arguments for oxygen (too detailed to go into here) being more believable in general.

"We can look at the composition of iron meteorites as well, which are remnants of small planetary bodies from early in our solar-system's history that segregated small cores. The compositions of these metal alloys match closely what we predict the composition of our core is using the evidence discussed above." [22] (See picture bellow of the anatomy of the earth)

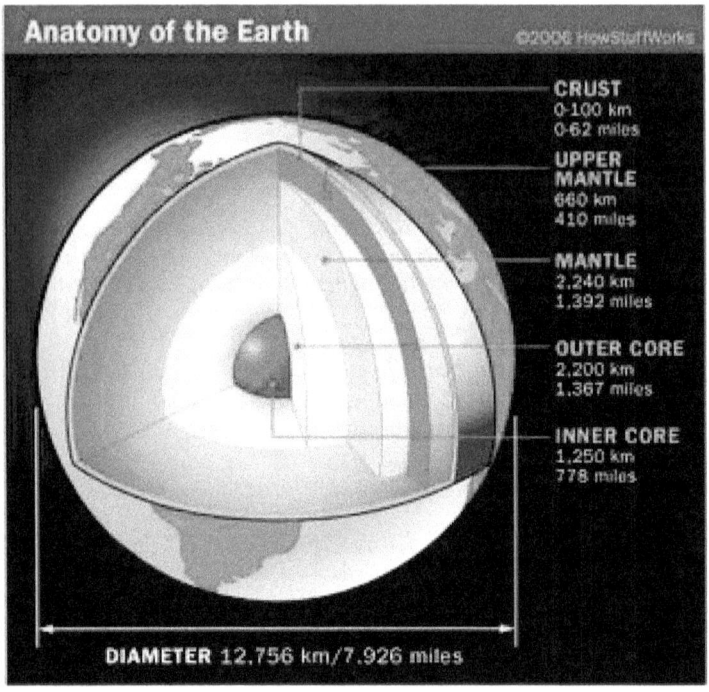

Picture [21]

"The interior of the Earth, as one of many telluric planets, is stratified, which means it is organized in superimposed concentric layers, having increasing densities as we go deep into the center of the earth...Its various layers distinguish themselves by their petrological nature (chemical and mineralogical contrasts) and their physical properties (changes of physical state, rheological properties)". [21B] (Translated from French)

MATHEMATICS – CHEMISTRY IN THE PYRAMID

Having closely examined the Pyramid of Giza using scientific methods, scientists have universally been astonished by the accuracy of the measurements of the structure, as illustrated by data from the following research (published on Wikipedia). According to the work of the Egyptologist Jean-Philippe Lauer, in his book: "Les Mystères des Pyramides - The Mystery of the Pyramids," (1988),"—in the page chapter-"Mathematical Observation of the Pyramid of Giza (Gizeh)," the following analyses have been observed:

"*When one studies the geometry of the large pyramid, it is difficult to make a distinction between the intentions of the builders and the properties that derive from the proportions of the building. The golden number and the number pi are often used to explain the proportions of the Pyramid: the Egyptians have, as we saw, built a slope for each face that can be expressed as 14/11.*

"*Concerning the golden number, the proportion of 14/11 entails an apothem/half-bottom report equal to*:

$$\frac{\sqrt{14^2 + 11^2}}{11} \simeq 1,61859$$

close to: $\quad \varphi = \frac{1 + \sqrt{5}}{2} \simeq 1,61803$

"*The value of the number would be expressed in the relationship (half-perimeter of the base)/height, resulting in the approximate value*:

$$\frac{4 * 11}{14} = \frac{22}{7} \simeq 3,14285 \simeq \pi$$

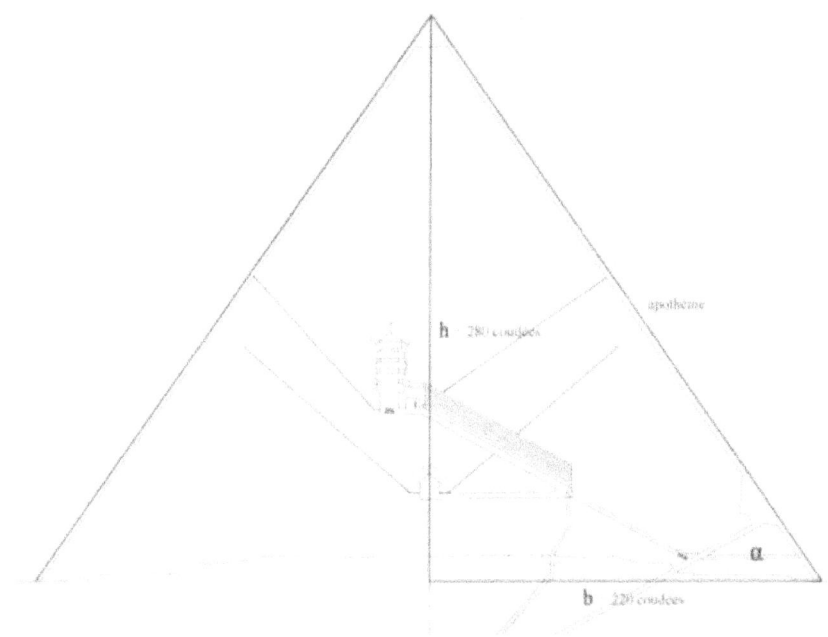

"*These two results derive from the use of a slope equal to 14/11. If a deliberate choice were made to write these numbers down during construction, the merit would be due to the architect who first used this slope in the Pyramid of Meïdoum, completed under the reign of Snéfrou. But this proposition is hardly plausible. According to a few rare mathematical documents in collections today, the Egyptians of antiquity had no knowledge of the number, and used another number expressed as 256 / 81 = 3,1605 to calculate the area of a sphere?*" (See Geometry in

ancient Egypt) [22A]

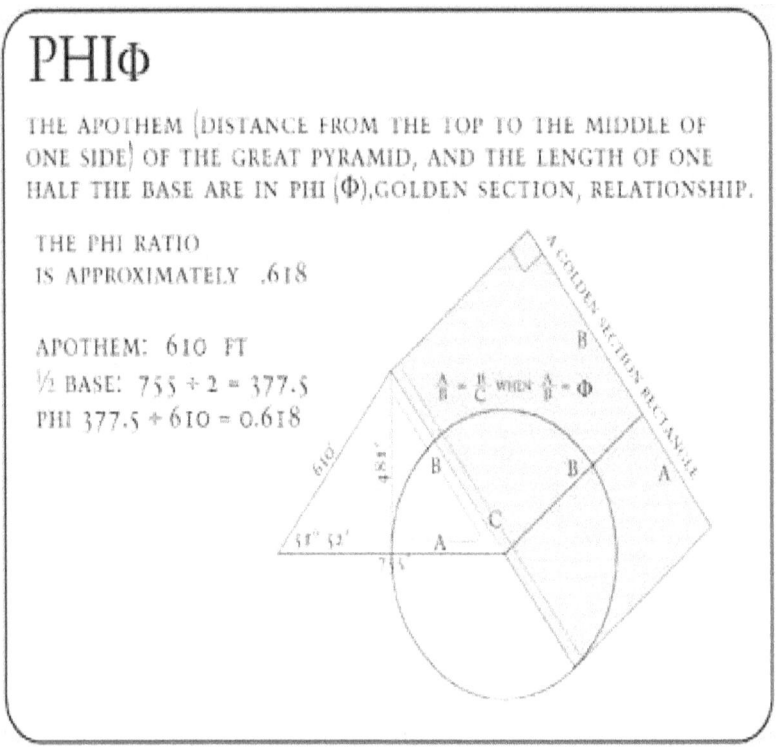

PHIФ

THE APOTHEM (DISTANCE FROM THE TOP TO THE MIDDLE OF
ONE SIDE) OF THE GREAT PYRAMID, AND THE LENGTH OF ONE
HALF THE BASE ARE IN PHI (Φ),GOLDEN SECTION, RELATIONSHIP.

THE PHI RATIO
IS APPROXIMATELY .618

APOTHEM: 610 FT
½ BASE: 755 ÷ 2 = 377.5
PHI 377.5 ÷ 610 = 0.618

Picture [23]

Here is another good question to ask: Why did the builders measure and create with such accuracy the capacity of the King's Chamber by applying the Pythagorean Theorem (Hypotenuse^2 =Adjacent^2 +Opposite^2; or $C^2=A^2+B^2$), as illustrated in the following image?

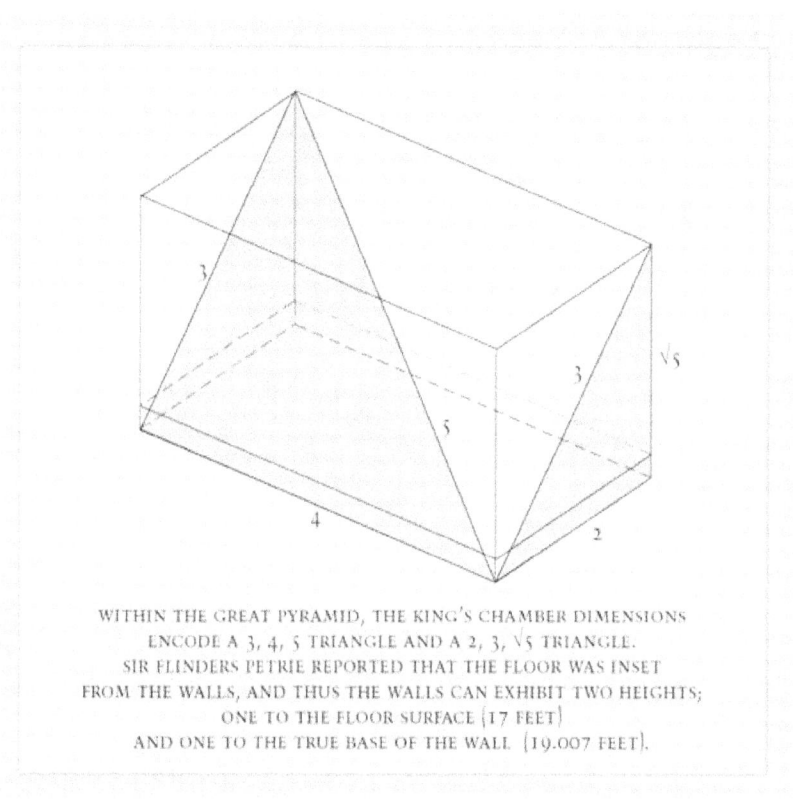

WITHIN THE GREAT PYRAMID, THE KING'S CHAMBER DIMENSIONS
ENCODE A 3, 4, 5 TRIANGLE AND A 2, 3, √5 TRIANGLE.
SIR FLINDERS PETRIE REPORTED THAT THE FLOOR WAS INSET
FROM THE WALLS, AND THUS THE WALLS CAN EXHIBIT TWO HEIGHTS;
ONE TO THE FLOOR SURFACE (17 FEET)
AND ONE TO THE TRUE BASE OF THE WALL (19.007 FEET).

Image [24]

(*Within the Great Pyramid, the King's Chamber's dimensions are coded A 3, 4, 5 triangle and A 2, 3, √5 triangle. Sir Flinders Petrie reported that the floor was set in from the walls, and thus the walls can be measured either from the surface of the floor (17 feet) or from the actual base of the wall (19.007 feet)*))

Let us verify the Pythagorean formulas in the King's Chamber for the first triangle: 3, 4, and 5

Hypotenuse =5; Adjacent side = 3; Opposite side = 4

25=9+16 ➔ 25 = 25

For the second triangle: 2, 3, √5

Hypotenuse = 3; Adjacent side = 2; Opposite side = √5

9=4+5 == =.> 9 = 9

And what can be said about the accuracy of the dimensions of the Queen's Chamber, as well as those of the descending and ascending passages, and the subterranean chambers?

For a long time this theorem was ascribed to Pythagoras; but in view of these small observations in the Pyramid, could one hypothesize that Pythagoras studied mathematics in Egypt? Since history tells us that Pythagoras was a student of the mysteries of Osiris and Isis, can we speculate that the Pythagorean Theorem first originated on the African continent, rather than in Greece as we are told in our school lessons and history books?

If Pythagoras was indeed the inventor of the theorem, it would mean that Pythagoras was born before the construction of the pyramids.

Also, why did the pyramid builders choose to make them from the hardest granite, rather than from other granites used elsewhere within the pyramid? Was it perhaps to contain a certain chemical reaction or gas? Perhaps the capacity of the coffin or sarcophagus represented the exact volume of chemical electrolytes or battery acid needed for each operation, based on the voltage capacity (energy) that the Pyramid generated for its full usage?

We have learned of the presence of those chemical products and water within the Pyramid of Giza thanks to the careful observations of Christopher Dunn. But Egypt is not the only place where this possibility presents itself. Inside some of the ancient pyramids of China, scientists have discovered the presence of mercury. Mercury is known as a good conductor of electricity.

Just as batteries are all made differently with different voltage capacities, so are the pyramids. Some of them do not have stress-relieving chambers, while others do. The Pyramid of Giza does have this type of chamber.

If batteries are designed differently to store certain levels or amounts of electrical voltage or power, then the pyramids too may have been designed to produce or provide different voltages or power. Those with a higher voltage and capacity can be identified by the presence of the stress-relieving chambers, the horizontal granite beams and the supporting limestone beams above the King's Chamber, as well the presence of an ascending passage.

The Maidum and the Darshur Pyramids of Sneru in Egypt and elsewhere, which do not have a stress-relieving chamber or an ascending passage, must have been designed for a lower voltage and capacity.

Here is the important message: The builders apparently knew a great deal about the effects of chemical reactions on stones/rocks composed of certain concentrations of minerals. When the pyramid is exposed to cosmic energy, especially the energy of the sun, at a certain height, a desired electrical effect is produced resulting in a transformation of energy.

Thus, the builders manufactured the pyramids by using knowledge of a chemistry unknown to us, including the combination of different minerals (volcanic rocks, stones, etc.) to produce a material (stone) having the right chemical capacities to allow it to capture solar energy as naturally as possible and to transform it into usable energy. The transformation takes place through a second process during a stage of the processes leading to the Pyramid of Giza.

It appears to be an effect that our contemporary scientific community has not yet discovered. When the construction of the Pyramid of Giza was completed, an element, which I will not name here, was placed at the top, or apex. This element acts as a receiver of light, or energy, from the sun, while also acting as an amplifier. Under the Law of Polarity, it receives rays or waves of light energy from the sun and those waves are naturally drawn down into the Queen's Chamber, where the same element of opposite polarity is placed, playing an important energy-amplifying role within the pyramid.

Once the final steps of creating the battery within the pyramid are completed and loaded up with battery acid (probably a mixture of sulfuric acid and water, as evidence points out), the pyramid is then naturally powered by electricity created through contact of light waves entering from the element at the apex of the pyramid, with the granite beams located in the stress-relieving chamber.

ENERGY FROM THE SUN

I believe that when the cosmic energy and the energy from the sun make contact with the horizontal granite beams in the stress-relief chamber [picture 24A], the solar or cosmic energy is converted into electricity. This works due to the chemical properties of the granite. Note that there is a second step in the process…

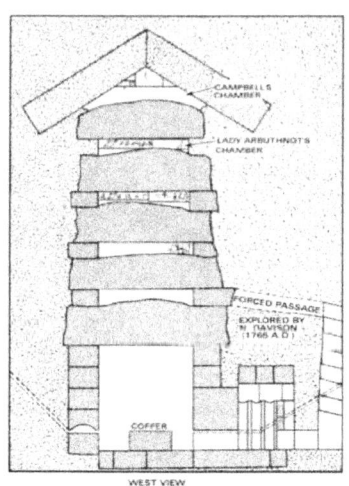

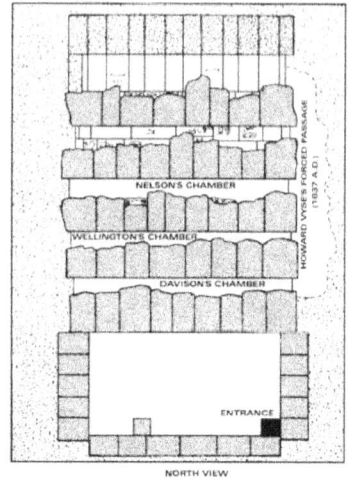

WEST VIEW

NORTH VIEW

RELIEF CHAMBERS ABOVE KING'S CHAMBER

GRANITE
LIMESTONE

Ralph Lyman

Picture [24A]

The Great Pyramid of Giza Association, published an article entitled "An Arab Who Got the Shock of His Life on the Summit," which may help to explain this electricity-generating phenomenon.

"Sir Siemens, a British inventor, climbed to the top of the Pyramid of Giza with his Arab guides. One of his guides called attention to the fact that when he raised his hand with outspread fingers, he would hear an acute ringing noise. Siemens raised his index finger and felt a distinct prickling sensation. He also received an electrical shock when he tried to drink from a bottle of wine that he had brought with him... When he held it above his head, it became charged with electricity. Sparks were emitted from the bottle..." [25]

HIGHER KNOWLEDGE OF GEOLOGY

Why would the builders of the Pyramid of Giza choose granite for use as building blocks? Perhaps they knew something about the chemical properties of granite that our scientists have yet to realize.

Let's look a little further into the chemical properties of granite.

According to the United States Geological Survey (USGS), *"granite contains a high presence of silicon dioxide (SiO2), and is composed of at least 65% of silica. The key silica elements comprising granite are quartz, feldspar and mica; and the dominant element in granite is quartz. Because of the presence of quartz in granite, it is likely subject to the quartz electrical property of piezoelectric, pyroelectric and may be tribolumiscent."* [26]

Pierre and Marie Curie from France, were   pioneers who realized the electronic potential or "piezo" of quartz crystal. As mentioned by Frank Dorland in his book "Holy Ice:"

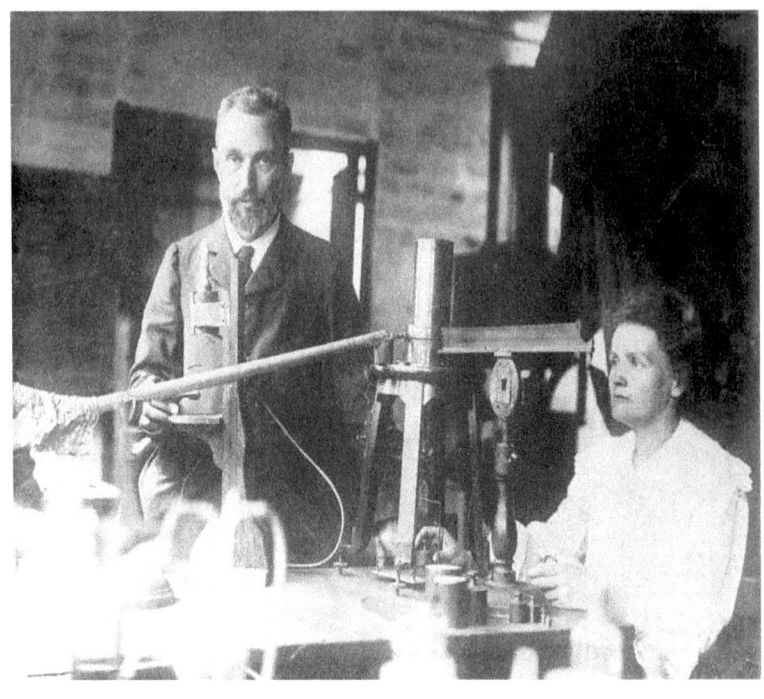

Image of Pierre and Marie Currie in their laboratory [27]

"The Crystal generates electricity when pressure is applied and displays polarity due to its positive and negative forces. Also, electricity fed into quartz crystal generates stress, which usually results in vibrations and precise oscillation... Like a microphone, the crystal is merely a responsive electronic device that reacts to energy (heat as well?!) and vibrations."[27A]

When the sun's energy makes contact with the element at the apex of the pyramid and moves down through the granite beams of the relief chambers, a natural form of energy is created and then converted into electricity, providing a continuous means of powering the battery.

The builders must have also built higher voltage and higher capacity batteries similar to the pyramid of Giza in different locations on our planet to serve the same function, because the rotation of the Earth around the sun brings different seasons into play: summer, fall, winter and spring.

It should be no surprise to our contemporary civilization that pyramids have been built all over the world: this information has been made public due to the efforts and courage of Dr. Sam Osmanagich, who discovered the pyramids of Bosnia in Europe; He is the author of the book "*Pyramids Around the World.* ". According to his investigations, there are pyramids in the U.S.A (250), on the Island of Mauritius as well as on the Canary Islands, etc.

Furthermore, according to investigations made public in an intriguing 21st century documentary entitled: "**The Revelation of the pyramids,**" the author / narrator, Jacques Grimault, and director Patrice Pooyard, raise the following question: *why are most pyramids located on a straight line that makes a 30 degree* *angle in relation to the equator? Is this mere coincidence?!*

Answer: No. Because we all know that it is hotter at the equator than at the North and South poles of planet Earth.

Are there other proofs relating to the components of the battery inside the pyramid of Giza?

After carefully comparing several research papers written by pioneering scientists long ago as well as in recent years, I found additional evidence relating to the components of the battery within the pyramid of Giza. The work of Christopher Dunn is available in an online article entitled *"Evidence of Ancient Electrical Devices Found in the Great pyramid?"*

Dunn and his team undertook scientific research for which I have deep respect because it is the result of efforts and material sacrifices like mine and those of Jacques Grimault and Patrice Pooyard, as well as the efforts of so many other researchers.

Another prominent pyramid researcher whose work requires close attention is the Puerto Rican, Samuel Laboy, a former civil engineer for the United States Army. Here is a short prologue from his book titled: *"A Civil Engineer Looks at the*

Great pyramid," by a professional engineer, Eduardo Ramirez, Ph.D:

"The book inspires and illustrates the full analytical study of the geometric planning and design of the Great pyramid of Egypt, which resulted in the accomplishment of the monumental work of Cheops' pyramid, marvel of the World in the past, in the present and in the future.

In this study the author took into consideration both the external as well as the internal configuration of this Monument, which is considered evidence of the greatness of the Mathematicians and the Civil Engineers of the old Egyptian epoch.

This comprehensive work describes, with luxury of geometric details, the analytical method, which defines the dimensional characteristics that the Great pyramid possesses, both in its exterior aspect as well as in its interior geometric details.

It is amazing that by means of this analytical geometric method you can corroborate with great precision all dimensions and particularities of the prominent engineering structure, which induces the intellect to contemplate the Egyptian Marvel.

The Cheops' Square pyramid consists of about 2,300,000 limestone blocks of an average weight of 2½ tons and reaching a maximum weight of 15 tons. These data, in part, were the motivation for this geometrical engineering study of the Great pyramid. The design and construction of this monumental structure lead the author to dedicate his intellectual and engineering knowledge to the presentation of his magnificent and excellent work.

Only one facet of man's evolving culture has exhibited virtually uninterrupted progress: the ability to convert to his use, materials present in the environment so that his reach was extended, his strength multiplied, and his work time shortened. This ability to create and use his intellect is man's technology. Congratulations to the Egyptian Mathematicians and Engineers!

I recommend this interesting and invaluable reference book to all scientists, mathematicians and engineers interested in the Egyptian pyramids."[27B]

In order to grasp the scientific finality of the pyramids, I invite you to view the following video documentaries on DVD or

YouTube: "The Revelation of the pyramids," "Inside the Great pyramid of Egypt 3D Computer Graphics," "Virtual Tour Through the Great pyramid for Scientific Research ,Purpose," as well as "The Forbidden Archeology," which present proven facts. They also are available on our website.

I invite you to browse Mr. Dunn's web page [28] to learn more on your own.

I believe that Mr. Dunn and his team, as well as those of "'The Revelation of the pyramids" deserve a lot of attention and applause for their work, considering the numerous challenges they have confronted until today in the face of a skeptical public (politicians defending their own interests rather than those of the humanity, scientists with fixed ideas, religious fanatics, etc.).

These authors attempt to pass on the results of their findings concerning the deeper questions that might reveal the setting of an ancient scientific revolution: to bring help to our civilization and a positive inheritance to future generations, an invitation to the evolution of the conscience of humanity.

I also understand these challenges: since more than once in the United States, I met resistance to this message, at the political level as well as in the media, in addition to facing religious intolerance. The precious message that we try to transmit in this book is intended for the benefit of future generations, including people of all races and all religions.

Yet our contemporary civilization has not yet unlocked all the knowledge of geology used by the pyramid builders. Unfortunately, the few scientists who have been able to reveal some of the secrets of converting cosmic energy from certain igneous rocks into electricity have been discouraged.

resonant
quartzite
chamber

We, as a society, fail to encourage people to continue to prosper in their fields of study and to master this knowledge of using free energy for the survival of mankind. And this is probably done deliberately to support the use of other energy resources. But given the solar storms crossing our weak magnetic field, we already know and have experienced solar winds that have disrupted our power grid. The result is that millions of homes, businesses and infrastructures risk a loss of electricity, especially hospitals in developed countries (where they do not possess back-up generators). And this should be a concern for Europe as well as other continents.

Do we have to wait for more dramatic consequences until we are eventually forced to regress as a civilization? This will happen if we do not act now by putting more resources into supporting research and development for the advancement of free, cosmic energy technologies.

ELECTROMAGNETIC INDUCTION

Now I want to change the subject a bit. A brief introduction to electromagnetic induction will be helpful here.

In my research, I came across a discovery made by Colonel Vyse in 1836 inside the pyramid of Giza [28]. He found a flat iron plate (apparently a meteor), about 12"x4" and 1/8" thick, which seems to indicate an application of the law of electromagnetic induction, first defined by Michael Faraday.

What does this law state? John Meurig Thomas, FRS, in his book entitled "Michael Faraday and the Royal Institution (The Genius of Man and Place)," explains:

"The principle of electromagnetic induction: Faraday discovered that when a coil on one side of a soft-iron ring was either connected to or disconnected from a battery, an electric current passed through the coil on the opposite side of the ring..."[29]

If this principle is proved to be true, then why would the

builders of the pyramids create an electrical current within the pyramid?

Picture [30]

Would there be another meaning to this message written by Faraday for us, when he stated many years ago:

"Electricity is often called wonderful, beautiful; but it is so only in common with the other forces of nature. The beauty of electricity or of any other force is not that the power is mysterious, and Electricity is often called wonderful, beautiful;

but it is so only in common with the other forces of nature. The beauty of electricity or of any other force is not that the power is mysterious, and unexpected, touching every sense at unawares in turn, but that it is under law, and that the taught intellect can even govern it largely. The human mind is placed above, and not beneath it, and it is in such a point of view that the mental education afforded by science is rendered super-eminent in dignity, in practical application and utility; for by enabling the mind to apply the natural power through law, it conveys the gifts of God to man."

—Michael Faraday, *Notes for a Friday Discourse at the Royal Institution (1858). [31]*

ELECTRICAL CURRENT

Most of us in the modern world use electronics either at home or at work. We also know that those electronic devices need electricity to function. And in view of what we have discovered so far about the pyramids being used as batteries, the question we may next ask is, "Why would the builders of the pyramids seek to create electrical generator systems all over the globe?"

Well, maybe like me, you thought of the pyramids as huge combinations of granite and limestone piled very precisely one on top of another, a burial site for a dead king whose body was never found, and now a place for religious superstition and entertainment.

Most people don't know that there are pyramids located in places other than Egypt; for example, in African Sudan researchers have revealed the existence of more than 350 pyramids, built by the monarchs, when the country was probably a province of ancient Egypt and later, when it became the independent Nubian Empire. And recently, 35 pyramids have been discovered in north Sudan (in ancient Nubia).

"All are of reddish bricks, but some distinguish themselves by a particular architecture. Surrounding an internal dome, some constructions organize themselves according to a disposition that evokes the gardens to the French, a characteristic of the site of Sedeinga. Such a structure also meets on the site of Meroe, also in Nubia. According to Claude Rilly and Vincent Francigny, that frame these excavations of the SFDAS, the Koushses were extensively inspired by the Egyptian people. "[31A] (From French translation)

Image [31B]

THE MEXICAN PYRAMIDS OF THE SUN AND THE MOON

Can we find proof that energy was used in the pyramid of the Sun? Maybe a return to the archives of former explorers will tell us more about that.

In 1906, when the pyramid of the Sun was being restored, a large mineral tile made of mica was discovered in one of the top rooms of the pyramid. After further study, the mineral proved to be a type of black mica that came from Brazil.

And still more astonishing was the discovery of a 'Temple Mica' within 300 meters of the pyramid, in which two thick films of mica, of 27.43 meters square, were superimposed one over the other and placed on a platform of paved rocks.

Was it simply luck to have found this mineral from Brazil in the pyramid where it was kept merely for exhibit, or was it intended for use of a more scientific nature? To get to the heart of it, let's consult a dictionary of minerals:

"Mica is the name of a family of minerals, of the silicate subgroup of phyllosilicates formed mainly of aluminum silicate and potassium. Along with quartz and feldspar, it is one of the constituents of granite…

Mica is used as a dielectric (electric insulator) in high tension and high frequency capacitors. "[31C]

Now let's discuss the mathematical construction of the Mexican pyramids. In a report on the work at the site titled: 'Ancients - Wisdom', the author states: "The pyramid of the Sun in Teotihuacan (Mexico), presents the same basic dimensions and half of the height of the 'Large' pyramid in Gizeh... This means that the pyramid of the Sun incorporates 'Pi' the following way:

$(4x\ \Pi)$ = x h Perimeter / Circumference of the basis. "[31D]

Consequently, given the previous data, since the mathematical formula of Pi is also found in Mexico, can we deduct that the same engineers or the same civilization (that I take the liberty to name here) 'Atlantis', could be responsible for the technological origins of the pyramids in Egypt? Could Atlantis be responsible for the construction of pyramids...on the whole planet?

Pyramid of the Sun in Mexico

In China, thanks to the careful observations of an American pilot at the end of the Second World War (1945); and thanks to the curiosity of some local Chinese residents, and in spite of the government's determination to prevent the international community from knowing about these structures, we are now aware of their existence: Among them is the pyramid of Maoling: see photo below.

The picture shows the existence of a pyramid in a faraway region of China: in the valley of Ya-sen, close to the city of Xi'an, a former capital. Curiously, like all the pyramids of the plateau of Gizeh, they are oriented to correspond to the four cardinal points.

Is this a simple coincidence, or the result of scientific calculation?

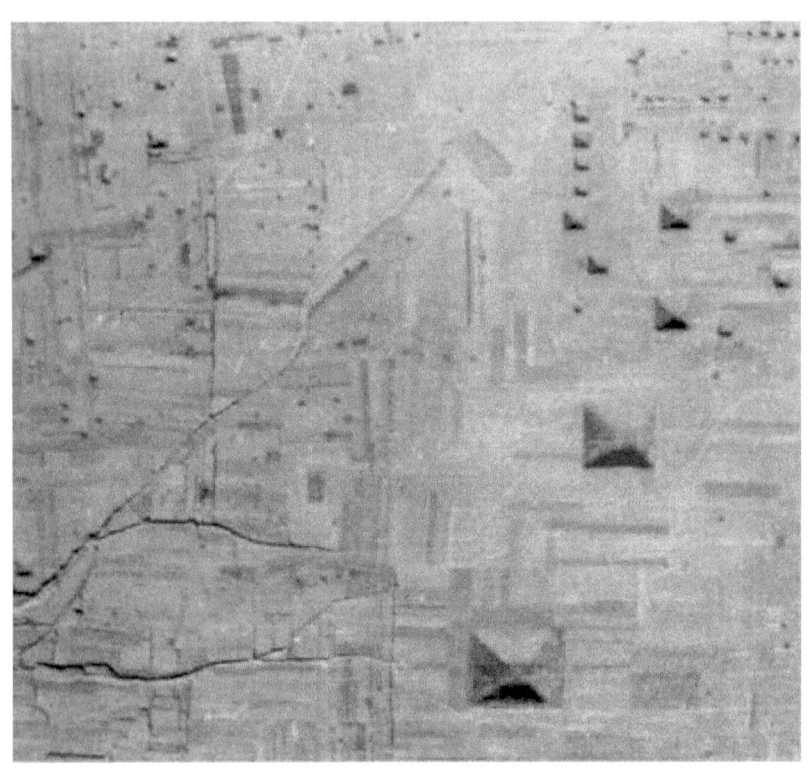

Image of the complex pyramids of China [31E]

EUROPEAN PYRAMIDS

In Italy: the Pyramid of Cestius (Pyramide di Caio Cestio or Pyramid of Cestia).

Elsewhere in Europe: the archaeological research posted on the internet site "Bosnia pyramids.com," informs us of the followings:

Dr. Osmanagich published the work: **"Pyramids Around the World & Lost Pyramids of Bosnia"**. *In his research, he proposes that the workers who created these "constructions" carved the hill in the shape of pyramid before covering it with a sort of primitive concrete. The largest hill, on which the works have been done, is about 70 meters high. Its base is a square whose sides measure 220 meters. After the discovery in August 2005, with the help of a probe, "numerous soil anomalies" were found to a depth of 17 meters. [Dr. Osmanagich] returned accompanied by experts to further his research. A geologist on the site, Nadja Nukic, explained the discovery of three layers of an unknown, polished, brownish stone, placed at an equal distance... In the Bosna Valley (Bosnia) in the village of Ozimi there is a strange major concentration of stone balls that can be found through Bosnia. The biggest of these balls measures 1.7m of top* *with a circumference of 5.3m. Most of these balls are concentrated on a hill in the region... " [31F].*

This pyramid, along with others discovered in the region, is under exploration, and is open to the public. You can learn more about the research on these pyramids by visiting "The Bosnian Pyramid.org."

Image of the pyramids of Bosnia

Even though several pyramids have been recently discovered, we submit a question that calls for a scientific answer: why would those precisely carved and positioned blocks of stone be placed where they are, circling the globe from Africa to China, to South America, to Europe and elsewhere? Do you think they were built with so much effort and expense, some of them precisely oriented toward the four cardinal points, without scientific or critical meaning and no logical purpose for the lives of the inhabitants of all these locations?

Let's start again with the battery explanation. We all know that a car needs a battery to power up its engine so that we can use it.

But all these immense, immobile structures, which were built before the concept of electricity was even understood (to our knowledge) -- their existence and construction is still unexplained. We still haven't figured out how they were built, without gigantic machines to hew and lay the enormous stones with such great precision.

Well then, let's ask the question I am so eager to have answered: Why would an intelligent people from a sophisticated but now dead civilization come up with a sophisticated design of a battery and create multiple batteries via these pyramids, and place them so precisely in a particular formation around the globe?

If my research in pursuit of an answer to my dream

intuition is accurate, we have learned that the pyramids are an advanced, highly sophisticated type of battery, apparently created for the purpose of producing an electrical current within the pyramid. See the following illustration:

On to the next question: What happens during the electrical process and why did the builders need to construct such mysterious forms within the pyramids? The answer is close at hand.

Hans Christian Oersted, a Danish professor of physics and chemistry, discovered that an electrical current generates a magnetic field around it. As reported by Mrs. Gillian Turner, in her book entitled "North Pole, South Pole."

"Hans was giving a demonstration to a group of students during a lecture at his home in Copenhagen...he was trying to show his students that electricity and magnetism were unrelated phenomena. What happened, though, was that when Orsted held a wire carrying an electric current over a compass, the compass needle swung around until it was at right angles to the wire. When the current was northward, the compass needle also reversed: a southward current made it point east." [34]

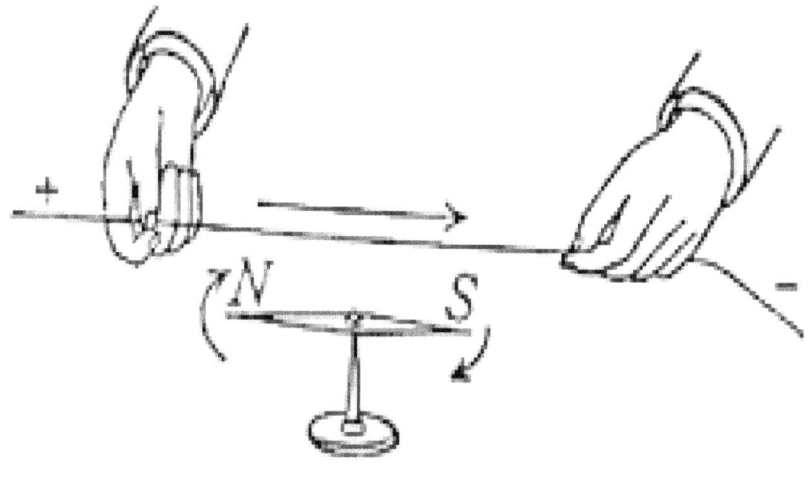

Picture [36]

Rick Groleau from Nova further explains:

"When an electric current passes through a metal wire, a magnetic field forms around that wire. This is the basic principle that allows electric motors and generators to operate." [35]

Image of Hans Christian [32]

"In that memorable year, 1822, (Hans Christian) Oersted, a Danish physicist, held in his hands a piece of copper wire, joined by its extremities to the two poles of a Volta pile. On his table was a magnetized needle on its pivot, and he suddenly saw (by chance you will say, but chance only favors the mind which is

prepared) the needle move and take up a position quite different from the one assigned to it by terrestrial magnetism. A wire carrying an electric current deviates a magnetized needle from its position. That, gentlemen, was the birth of the modern telegraph."

— Louis Pasteur, *in the Inaugural Address as newly appointed Professor and Dean (Sep 1854) at the opening of the new Faculté des Sciences at Lille (7 Dec 1854), in René Vallery-Radot, The Life of Pasteur, translated by Mrs. R. L. Devonshire (1919) [33]*

If a civilization of such intelligence could imagine a natural design, to create a battery using only rock formed in the Earth's interior, then wisdom and curiosity might compel us to ask yet another question: Why did they choose to use only rocks and minerals formed inside the Earth? Well, I can only guess here that it was possibly to resolve a problem that also originates deep in the Earth. Groleau adds:

"In the Earth, the liquid metal that makes up the outer core passes through a magnetic field, which causes an electric current to flow within the liquid metal. The electric current, in turn, creates its own magnetic field, one that is stronger than the field that created it in the first place. As liquid metal passes through the stronger field, more current flows, which increases the

field still further. This self-sustaining loop is known as the geomagnetic dynamo."

Other ideas emerged to attempt to explain the scientific phenomenon of generator geomagnetism that takes place inside the earth. On this topic, we are going to have to resort to an eminent professor and specialist in the field: Professor J. Mervin Herndon, PhD in Geosciences, and author of more than four works of scientific research on the topic of the internal structure of the Earth. Here is an excerpt from one of his articles (2007), where Herndon presents the obvious proof:

"The georeactor sub-shell being a slurry or fluid, he suggested that the Earth's magnetic field is produced by a dynamo-mechanism operating in the georeactor sub-shell [9]. Significantly, within the georeactor sub-shell there is no impediment to long-term, sustained convection; heat generated by nuclear fission in the sub-core causes the fluid at the bottom of the sub-shell to be lighter, more buoyant, making it rise to the top of the sub-shell, where it contacts the relatively good thermal conductor heat-sink that is the inner core, which is in contact with another relatively good thermal conductor heat-sink, Earth's fluid core. Thus, there is no impediment to long-term, sustained

convection in the georeactor sub-shell [9-11]."Cf. the above picture [37]

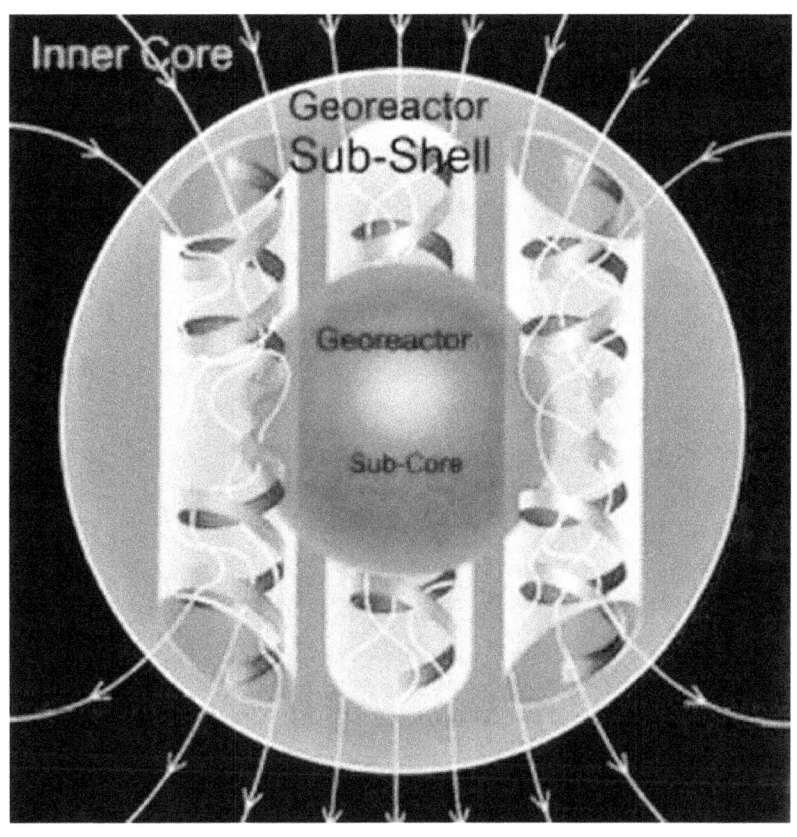

THE CIVIL ENGINEERING OPERATION OF THE PYRAMID OF GIZA

Without any scientific education on the topic, it can prove difficult to grasp all these explanations of generator geomagnetism and to have a logical understanding and to comprehend how the pyramids were used as batteries for bringing energy from inside the earth. While I was bewildered by this explanation of the processes between the pyramids and earth's interior, I remembered another dream, which illuminated my understanding while restoring my confidence.

The second dream I had that night clearly stated that the concept of the "septic tank" was also critical to the construction of the pyramids, and particularly to the pyramid of Giza.

In the dream, I seemed to be sitting in a classroom. I was being taught how wastes from the toilet moved to the septic tank underneath our house. I could not see the teacher's face, but while I listened to his explanation, I saw a visual demonstration.

When I awoke, I felt a little bit scared because it seemed so strange to me. Why was I being taught about a waste disposal system? By linking this dream to the first one I'd had that same night, when I was introduced to the concept of a battery with its plus and minus poles, my mind began to relax.

I let go of my fear, realizing after a short while that the God Force or Divine Spirit was actually revealing the second of two fundamental scientific purposes of the pyramids to me. It was easy for me to understand and realize the scientific concept of the battery as I did my research, but it took me a month to fully understand, absorb and realize the concept of the septic tank and how it is applied in the system of the pyramids.

To explain why the pyramids incorporated the civil engineering concept of the septic tank, the tank used in our house was used as an example.

I think perhaps it's best to share this information in the same way I received it, by explaining the function of the septic tank that is a part of all residential or commercial properties.

I must confess that I had some doubts on this aspect of the "civil engineering of the pyramid." For a time I thought that the septic tank in my own house was the only subject of the dream. I believed that I was being given a message to pass on to the owner of the house, so that she could make an inspection of her septic tank. After having this dream, I started believing that this house had many problems; I tried to understand how I had managed to live so long in this house that presented so many problems.

Let's pursue our adventure!

On July 04, 2011, when I moved from the state of Massachusetts to Washington State, I lived in the city of Everett, a few kilometers from the place where I had my "pyramid dreams." I didn't want to leave the state of Massachusetts that I liked so much, because I had just published my first book four months earlier; but I had a dream.

In this dream, a man appeared and invited me to take some walks with him, insisting that I move to Washington State. While we walked together like old friends, I could see the sunset on the horizon before us.

I liked living in the city of Boston and wanted to stay for a long time to market my book, but now someone, a messenger sent by providence, was wisely asking me to change my schedule; it was because of him that I was willing to do so.

While we spoke and walked, I felt that this man was steeped in scientific knowledge. He was indeed kind and cordial in his way of telling me that I should move to Washington State. His face seemed familiar to me; I had the impression of having seen him somewhere but I could not remember where.

Listening to him with a sense of submission, I promised to leave Boston for Washington State on the 4th of July. I had no idea why I was moving away. I simply followed my guidance. He was a messenger (Spiritual Master) sent by my spiritual guides (the Inner Master) to embark on an unknown, divine mission in the service of life.

Once I arrived in Washington State, my financial situation required me to adapt in order to survive; this is how I found a room to rent in return for helping a white woman who was about 70 years old. I liked neither the neighborhood nor her house, but I remembered a dream in which I lived with an aged woman who was in need of assistance; and since I was professionally trained to help elderly people in need, I decided to stay there in her house. And now I realized that the walk that led me to find that house was not a random event. I had become acquainted with this woman from a dream before meeting her in waking life; so the dream guided me to submit to living in the city of Snohomish, just one month after moving to the city of Everett in Washington State.

While the good woman tended to her activities – and as she also studied dreams -- I told her the dream of "the septic tank." She responded by explaining that her septic tank had been failing for years, and needed to be checked.

This discussion increased my curiosity about the workings of septic tanks. The woman asked me to look for simple solutions that would not involve a major expense to maintain her septic tank. So I began my investigation on this very topic and, inspired by my dream, I found an explanation in a text written by an expert in this field: the U.S. Department of Health, Education and Welfare. Here is how it is explained:

"Functions of Septic tanks:

Untreated liquid household wastes (sewage) will quickly clog all but the most porous gravel formations. The tank conditions sewage so that it may be more readily percolated into the subsoil of the ground. Thus, the important function of a septic tank is to provide protection for the absorption ability of the subsoil. Three functions take place within the tank to provide this protection.

Removal of Solids. Clogging of soil with tank effluent varies directly with the amount of suspended solids in the liquid. As sewage from a building sewer enters a septic tank, its rate of flow is reduced so that larger solids sink to the bottom or rise to the surface. These solids are retained in the tank, and the clarified effluent is discharged.

Biological Treatment. Solids and liquid in the tank are subjected to decomposition by bacterial and natural processes...

Sludge and Scum Storage. Sludge is a partially submerged mat of floating solids that may form at the surface of the fluid in the tank. Sludge, and scum to a lesser degree, will be discharged and compacted into a smaller volume...

If adequately designed, constructed, maintained, and operated, septic tanks are effective in accomplishing their purpose.

In brief, the liquid contents of the house sewer (a) are discharged first into the septic tank (b), and finally into the subsurface absorption field (c)." [38]

Conventional subsurface wastewater infiltration system

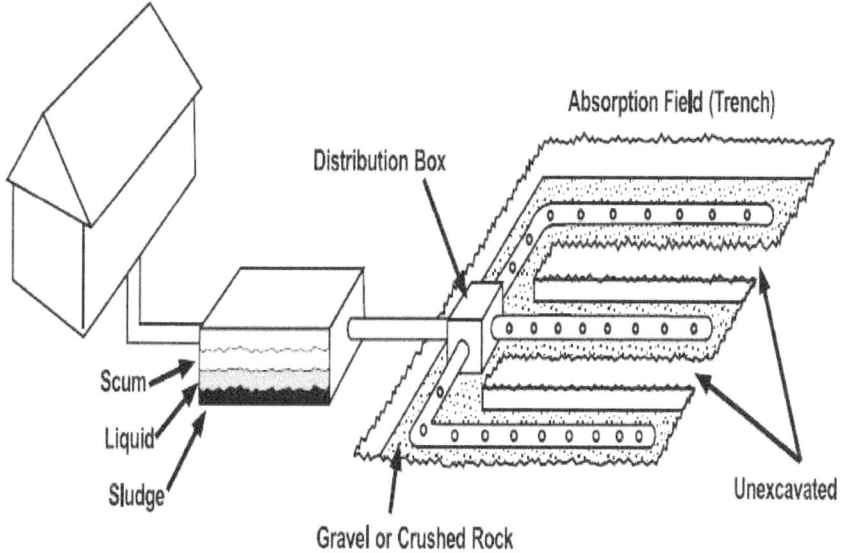

Picture [39]

While I knew little about septic tanks when I started my research, after reading and viewing materials related to this subject, I began to understand how the septic system works within the pyramid of Giza. After searching for audio and visual materials all over the Internet, only one gave me a comprehensive understanding that I could associate with the activity as it was explained in my dream, working through the pyramids.

This YouTube video is titled: "It's All Connected. An

overview of On-site Septic Systems," by Dig It Excavating Inc. [40]

Looking at interior photographs of the pyramids, I now understood that the presence of the King's Chamber, the Queen's Chamber, the grand gallery, the grotto, the first ascending passage, the granite plug down to the descending passage, the descending passage and the subterranean chamber are evidence of a functioning septic tank.

In the case of the pyramid of Giza, we are not dealing with the liquid contents from the house sewer. Instead, we are dealing with chemical products and treated water, the elements used to create a working battery. The designers of the pyramids must have been very intelligent to use this concept of the home septic system for eliminating waste in order to accomplish a desired result.

This points us in the direction of the ground beneath the Earth.

The process may also indicate that septic tanks were used in their homes, and they mastered that knowledge by creating pyramids or batteries.

This simple evidence leads a curious person to speculate that the pyramids were probably not designed by either the Egyptians or the Mayans. Here is an illustration of the septic tank diagram inside the pyramid:

Most pyramids were constructed close to water sources; (was it in order to facilitate the evacuations or distribution?!)

THE PYRAMIDS AND ELECTROMAGNETISM

The next questions I asked myself were the following:

What were the pyramid-builders trying to accomplish with this particular septic system?

Why would they waste their time playing with electrical currents and gas, sending them into the earth via a septic tank and gravity?

After researching and associating the scientific concepts of the battery with the septic tank, I could now understand how a chemical reaction could have been created in the King's and Queen's Chambers where the elements were combined, and how they were then separated through trenches from the grand gallery to the grotto, to the descending passage, finally reaching the subterranean chamber.

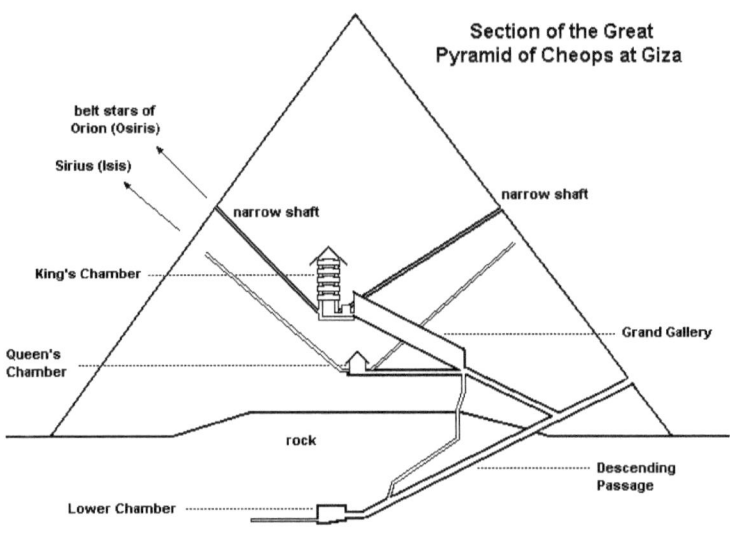

Picture [41]

However, I was still skeptical as to how electricity and magnetic waves could travel between rocks or stone to reach the earth's core, which seemed to be what the builders were attempting to do, based on my research. While reading "North Pole, South Pole," by Gillian Turner, I discovered the answer to these questions.

Her book taught me a lot about the science of electromagnetism, and I recommend it to any person interested in learning more about this topic. It was here that I discovered that electromagnetic waves can travel between solids, which explains why a magnet can attract iron from a certain distance through its magnetic waves, even crossing barriers such as a stone to reach the iron.

As to the law of electromagnetic propagation, Ms. Turner states:

"According to Maxwell's analysis, the speed of his electromagnetic waves depended on just two parameters, electrical permeability and magnetic permittivity, which respectively expressed the electrical and magnetic properties of free space... Maxwell was able to show that his predicted electromagnetic waves traveled at a speed indistinguishable from the best available measurements of the speed of light." [42]

I also gained some valuable information from a lecture on the speed of light. The lecture, entitled "Physics of Energy & the Environment," by Professor Dean Livelybrooks [43] tells us about electromagnetic waves.

Electromagnetic Waves (review)

Electromagnetic waves comprise the radiation we receive from the sun, and from each other. X-rays, visible light, microwave and radio waves are all examples of EM radiation.

These waves are characterized by:

Period (T) the time {seconds} between crests of the wave)

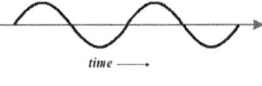

Frequency (f) the inverse of period-- 1/T {in 1/seconds}. How fast a wave is bouncing up and down. ("color" of wave).

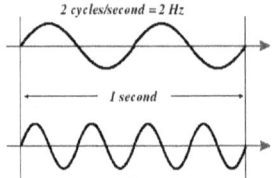

Wavelength (□) the distance {meters} between crests of the wave at a given time

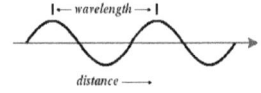

Wave Speed (v) the distance the crest of one wave moves in a second {meters/second}. The speed of most everyday EM waves is close to the speed of

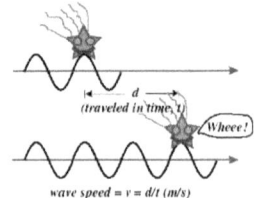

light, "c," approx.
300,000,000 m/s.

Ms. Turner's works came into my hands at just the right moment. The information they contained gave me more confidence about offering my ideas on the scientific purpose and function of the pyramids. My next challenge was to understand the ultimate scientific reasons for building the pyramids.

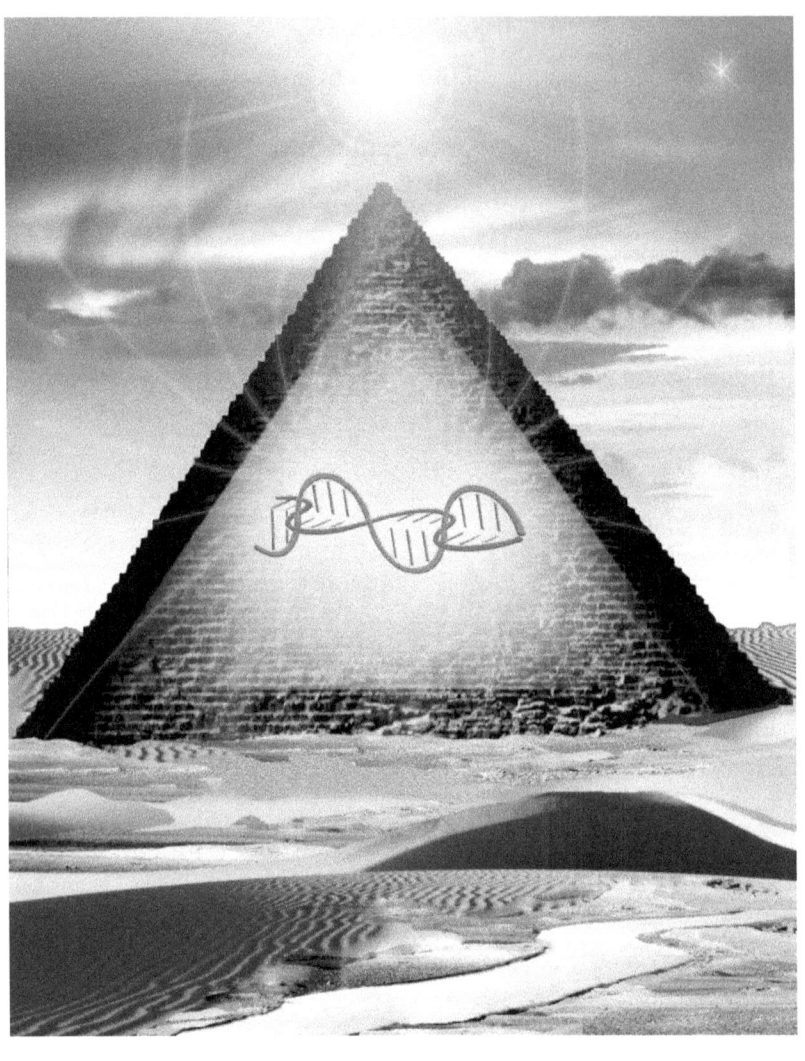

THE PYRAMIDS AND THE EARTH'S MAGNETISM

The more I read Ms. Turner's work on electromagnetism along with other investigations and research on the subject, the easier it became to link these findings with the situation we are facing in regards to the Earth's magnetic field.

The entire climate system is dependent upon the magnetic field. The biosphere maintains its balance through the existence and stability of the magnetic field. Animals use the magnetic field to find their way to their birthing grounds. Every living being is affected when the magnetic field does not function properly.

I discovered, for instance, that the Earth has its own protective shield to prevent infiltration of high radiation levels from the Sun and the galaxy. The existing magnetic field permits only certain energy frequencies or radiations from the Sun to reach the Earth, so that it is habitable.

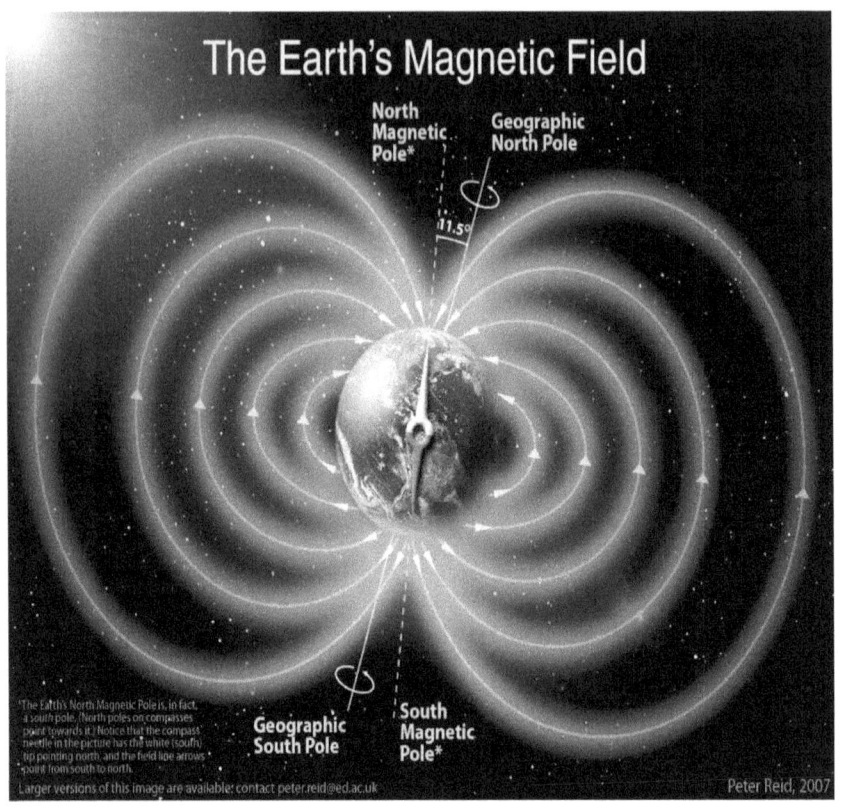

Picture [44]

"The terrestrial magnetic field that one can also call geomagnetic field, is an immense magnetic field that envelops the Earth, in a non circular way. To the extremities of the Earth, that is to say to the poles, it is the place where one finds the maximal intensity of the terrestrial magnetic field whereas he/it is much less intense midway between the poles."[44A] (From French translation)

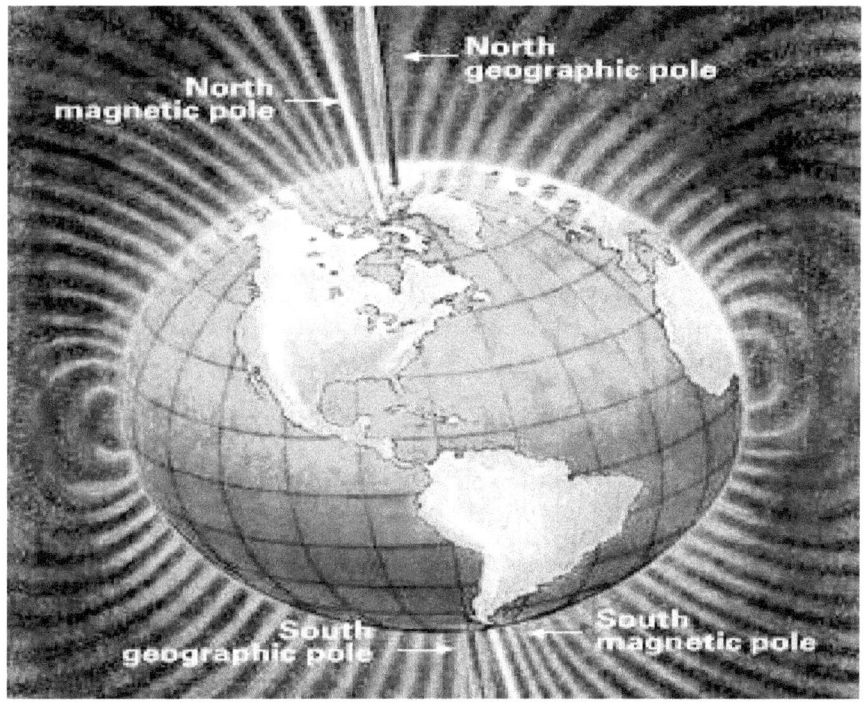

Picture [45]

"We live on a gigantic magnet: the Earth. It behaves like a magnetic dipole. Its magnetic axis, in movement, and its geographical rotating axis should not be confused [with each other]. "[44B] (from French translation)

As we learn in school, energy from the Sun fosters the growth of vegetation via photosynthesis. It also provides a directional function, which facilitates the migration of animals and keeps our water (oceans, rivers and lakes, etc.) from heating up to the point where it evaporates.

What would happen if all our sources of water dried up? What if our crops died, due to endless drought?

The magnetic field also maintains a balance between cold and heat, so that the ice in the Antarctic does not rapidly melt down, which could result in the oceans overflowing and dumping excess water into cities, towns and islands close to oceans and rivers, etc.

If you listen to or read the news, many stories like this are being reported around the world.

According to scientists, the central core of the Earth's magnetic field (or shield, to be more accurate), is located within our planet. Geologists call it "The Earth Core."

Based on the information given to me in my dreams, and what I confirmed through my research, this core is dominated by the presence of iron.

As I became aware of these things, the meaning of the necklace of pyramids "encircling" the earth at a 30 degree angle off the Equator, became clear to me. These pyramids were linked to the Earth's magnetic shield. I remembered that I'd had another dream prior to the two about the battery and the septic tank, and before the unexpected visit to the pyramid of Giza. Due to the timing of the dreams, I could not quite understand their relationship to one another, nor how to link them all to the pyramids and the Core (the Heart) of the Earth.

But when seen as a whole, the dreams began to make sense.

THE PYRAMIDS AND THE EARTH'S CORE

In that first dream, I survived an invasion of aliens in the form of machines made of iron. The aliens were chasing me and I was desperately running away from them to survive their deadly pursuit. The dream was a frightening one, like a scene from a science fiction movie.

A few days after having this dream, I went to the library in Snohomish, Washington to check out some books. On my way out of the building, after reading some online articles related to the pyramids, I also decided to check out some old-style movies to entertain myself over the weekend -- to relax. My attention was drawn to a film starring Tom Cruise, entitled "War of the Worlds." To my amazement, the movie (and the book upon which it is based) was quite similar to the subject of my dream. How could that be possible?! This coincidence gave me the chills.

But in spite of the similarities, I didn't make the connection to my dream, because I thought the dream was just a visit to a different planet or a place where I did not belong (a nightmare) while my physical body was asleep.

We all have dreams that sometimes don't make much sense when we are in the waking world. Here is a drawing depicting the dream.

Then, a week later, as if by magic, and without understanding the nightmare of "the machines," I had another dream of an iron machine. This time, I could see that the Earth was uninhabited and a strong heat was radiating from it. There were no living beings, not even a single tree. There was only a small mechanical iron machine that looked like a robot, but designed like a car with four wheels. It rolled by itself down an empty road. This small iron machine was now the master of the street. The dream made no sense to me at all.

However, a few months later when I started to learn about the core of the Earth being made of liquid iron, I was able to connect the dreams of the iron machines with the subject of the pyramids.

These two dreams were actually showing me that the revelation I received in my dreams regarding the pyramids was in fact an indicator of the existence of the Earth's magnetic core.

Following the sometimes odd and initially incomprehensible guidance of these dreams, I continued my education about the purpose of the Earth's core and its magnetic field. I became convinced that the element of iron, appearing in my dreams in the form of machines on the Earth's surface, is the primary constituent of our Earth's core, and that these machines were hungry, and in need of some kind of "food" of their own. The sun rays represented that energy.

If the iron within our Earth's core needs 'food', then I deduced that this food could possibly be electrical current or magnetic waves. I now began to understand what the builders of the pyramids had accomplished. I started to truly appreciate the scientific purpose of the pyramids.

While I was establishing my own understanding on the subject, I continued seeking scientific confirmation from other sources about this last revelation. And some time later I discovered the works of the previously mentioned Gillian Turner.

Guided by the God Force, I found myself at the Seattle library intending only to look for more books on the subject of the pyramids. It was in the library, as if by magic, that I saw her book; I ended up grabbing the book, which was sitting right in front of my nose on the display bookshelf, as if waiting for me.

Just its title "North Pole, South Pole" and the graphic of the planet Earth attracted me like a magnet and I exclaimed joyfully, "This cannot be possible!"

Surprised, seeing it right there in front of my nose on my way out of the seventh floor of the beautiful Seattle Library, I could not wait to check it out and read it.

As I read the first pages, I realized that indeed the God Force was continuing my education on the subject of the pyramids and their relationship to the Earth's magnetic field. I began to feel that my investigation and education on this topic were nearing completion.

The book was a summary of electromagnetic discoveries by many pioneers through to the present. I had been wondering how to explain the concept of the iron inside the Earth and its need for food (or energy), and my answer came through the following story:

"As early as 1831, Peter Barlow, an English mathematician and engineer, wrote a paper entitled "On the Probable Electric Origin of All the Phenomena of Terrestrial Magnetism."

Barlow had constructed a wooden globe with copper wires wrapped around lines of latitude. On passing a current through the wires he had produced a magnetic field similar in form to that which surrounds the Earth, and in doing so, demonstrated that Earth's magnetic field could be electrical in origin.

The question was: How might electrical currents be generated inside the Earth? Barlow suggested the heat of the sun was somehow transformed into electrical energy, but he was unable to come up with a credible mechanism." [46]

My God, I exclaimed in amazement, feeling light as a leaf in the wind, as though this simple paragraph was an intuition of the divine physical demonstration. I thought about Barlow, speaking to him as if he were at my side, thanking him for having begun the research that today reinforced my theory. Although he is no longer of this world, by the work of his mind, he felt like my divine savior!

Upon reading this paragraph, I felt satisfied that my spiritual understanding of the iron underneath the Earth looking for food made sense, and I wished that Mr. Peter Barlow could have been with me to enjoy this moment. Somehow, a person with no scientific aptitude or training had ended up understanding the answer to his question from a series of strange and confusing dreams.

The part of Peter Barlow's work, where he too suggested that the energy radiating from the sun was somehow transformed into electrical energy, reinforced my theory. I silently congratulated him for coming so close to the scientific truth.

As I read the book, I also learned that Mrs. Inge Lehmann was the first to theorize that the Earth has an iron-rich core. Again, this validated my inner revelations on this subject.

Regardless of these affirmations, I was still eager to learn more about a truth that I couldn't see but I could feel.

That truth within started with another of my seemingly endless list of questions: Why would this intelligent civilization build batteries (the pyramids) to feed the Earth's core with electrical current or magnetic waves, perhaps for centuries or millennia?

To answer this question, let us look out of the windows of our planetary home, and take a quick look at our closest neighbor, the planet Mars.

WHAT MIGHT HAVE HAPPENED TO MARS?

Over the past couple of decades, NASA has done tremendous work by sending robots to unlock the mysteries of what may have happened to the fabled Red Planet, Mars. A 2011 article by Nola Taylor Redd states:

"Astronomers have found more evidence that Mars was wet and warm in the ancient past... Water-carved landforms on Mars are just one source of evidence that liquid once existed on the planet... During warmer seasons, or after surface heating activity such as volcanism or a large meteor impact, ice could have melted and rushed across the land, cutting wide swatches. Once started, the racing torrents would have been difficult to freeze until they slowly tapered out..." [47]

How long have humans asked this tantalizing question? Perhaps you have wondered yourself. Did life exist on Mars?

On August 9, 2011, an article entitled "NASA: DNA

Found on Meteorites Indicates Life May Have Originated in Space," published in the Proceedings of National Academy of Sciences confirmed the following:

"NASA researchers have found the building blocks for life on Earth in meteorites, indicating that the components for life on Earth may have originated in outer space... the scientists found that ready-made parts could have crashed to the Earth's surface on objects like meteorites and assembled under earth's early conditions to create the first DNA... The team also found hypoxanthine and xanthine, which are not part of DNA but are used in various biological processes...

"People have been discovering components of DNA in meteorites since the 1960s, but researchers were unsure whether they were really created in space, or if instead they came from contamination by terrestrial life," said Dr. Michael Callahan, lead researcher of the discovery. "For the first time, we have three lines of evidence that together give us confidence these DNA building blocks actually were created in space."[48]

If there was once life on Mars, then what happened to it? We can speculate that Mars may have also had a magnetic shield of its own. Perhaps, we can anticipate the end of the story, supported by our explorations on the dry, dusty Red Planet, by theorizing that Mars' magnetic field may have disappeared, leaving it vulnerable to high levels of radiation from the Sun and from outer space.

Gillian Turner asked some of these questions in her book:

"However, perhaps the most interesting bodies in the inner solar system are the moon and Mars. On both, surface rocks seem to be strongly and permanently magnetized. Portions of the Martian surface even seem to display magnetic barcode-alternating stripes of oppositely magnetized rocks reminiscent of Earth's seafloor. Mars no longer has an internal magnetic dynamo, but it may have had one in the past. Perhaps an initially molten core cooled and finally froze solid, switching off the Martian magnetic field in the process. If so, did Mars once experience both polarity reversals and plate tectonics?" [49]

If the magnetic field is responsible for protecting our planet, it is obvious that its slow and gradual weakening or reversal may eventually destroy the protection we currently enjoy against high radiation levels. This means that the intensity of the Earth's magnetic field is directly linked to our climate.

I am now going to make a rather astounding statement: the weakening of the magnetic field is, without a doubt, the cause for the climate change we hear about in daily news reports.

THE EARTH'S MAGNETIC FIELD AND CLIMATE CHANGE

Through the progression of my dreams, and the corroboration of scientific articles, could one deduct that the stronger our terrestrial magnetic shield (ozone layer), the safer life is on Planet Earth?!

Let's look at a gradual reversal scenario. If the reversal scenario is indeed responsible for our climate change, what would then be the effects of such a gradual decrease of the Earth's magnetic shield?

"The great bulk of the liquid outer core of the planet, of course, is made of molten iron," said Lewis Page of *The Register*, an environmental reporting magazine. *"That's just as well for us and all life on Earth. As the spinning blob of superhot melted metal we all live on top of generates a tremendous powerful magnetic field which keeps off all the plasma storms and cosmic rays and suchlike deadly space radiation, so that we aren't fried out of existence on a routine basis."* [50]

Thanks to Mr. Lewis for reminding us so simply of the reality of climate change. In view of those reports, instead of focusing on carbon dioxide as a cause of climate change, perhaps we can agree here that climate change is a direct consequence of the weakening of the Earth's magnetic field. Let us learn more about what would happen if the Earth's magnetic field gradually loses its strength. The answer is consistent with what climate researchers have been saying about the warming of Planet Earth.

There is more information available to support this theory. I found an on-line article entitled "Global warming 'confirmed' by independent study; The Earth's surface really is getting warmer, according to a new analysis by a U.S. scientific group set up in the wake of the 'Climategate' affair. This information is not coming from shady news sources. It was reported by Richard Black, environmental correspondent for BBC News, which is a trusted and reliable organization, especially in its scientific reporting.

"The Berkeley Earth Project has used new methods and some new data, but finds the same warming trend seen by groups such as the UK Met Office and NASA.

"...The Berkeley group says it has found evidence that changing sea temperatures in the north Atlantic may be a major

reason why the Earth's average temperature varies globally from year to year…Since the 1950s, the average temperature over land has increased by 1C, the group found."[51]

Why should the health of our planet be the concern of all nations? Why is it so important and urgent for our scientists to take note of this new direction in the scientific theory of the pyramids? If my dreams are accurate, and all the research I've done is consistent with my discoveries, it is important to equip a team of international researchers to test the validity of the scientific revelations of the pyramids, to the extent that climactic conditions allow it, with the goal of reducing the negative impacts which will follow from atmospheric changes to come.

There is my message.

Have we unconsciously, if gradually, reached such a state of dysfunction that we face certain destruction? Have we been purposely blinded to recognizing the causes and solutions, because of short-sighted greed, religious intolerance and overwhelmingly toxic political interests?

Can civilized people of all nations and faiths, in the name of humanity and indeed of survival, come together in a united global partnership to learn and master this new technology? I am guided from within my heart to make this book available to the scientists of our beloved planet, in hopes of minimizing the impact of the destruction facing us, in order to preserve the rich diversity of life on Planet Earth.

Perhaps those forgotten, highly evolved civilizations from the past were the wise ones who built the pyramids around the Earth for the purpose I uncovered in my dreams and research.

They obviously faced similar challenges during their time, and perhaps the ones who listened deeply had dreams that helped them to develop scientific solutions to protect the Earth. So here I am, a simple person with no scientific training, hoping that what I have been taught might result in actions today so that we too may live and follow in their footsteps, and so that future generations may also live and learn how to resolve life's mysteries.

The ancient, forgotten ones left us with a tremendous number of signs that have lasted for millennia, and if I had to venture a guess, to warn us about the problem of the electromagnetic field that they once faced and how they dealt with it.

The ancients in the Nazca desert of Southern Peru left us with a particular symbol carved on rock, pointing North and South. Let us look at the Nazca lines, more carefully. See image below.

THE PARACAS CHANDELIER

Pictures from the film documentary "The Revelation of The Pyramids" [51A]

Marshall Brian, the founder of HowStuffWorks, explains the basics of how a compass works:

"The reason why a compass works is [more] interesting. It turns out that you can think of the Earth as having a gigantic bar magnet buried inside. In order for the north end of the compass to point toward the North Pole, you have to assume that the buried bar magnet has its south end at the North Pole... If you think of the world this way, then you can see that the normal 'opposites attract' rule of magnets would cause the north end of the compass needle to point toward the south end of the buried bar magnet. So the compass points toward the North Pole."[52]

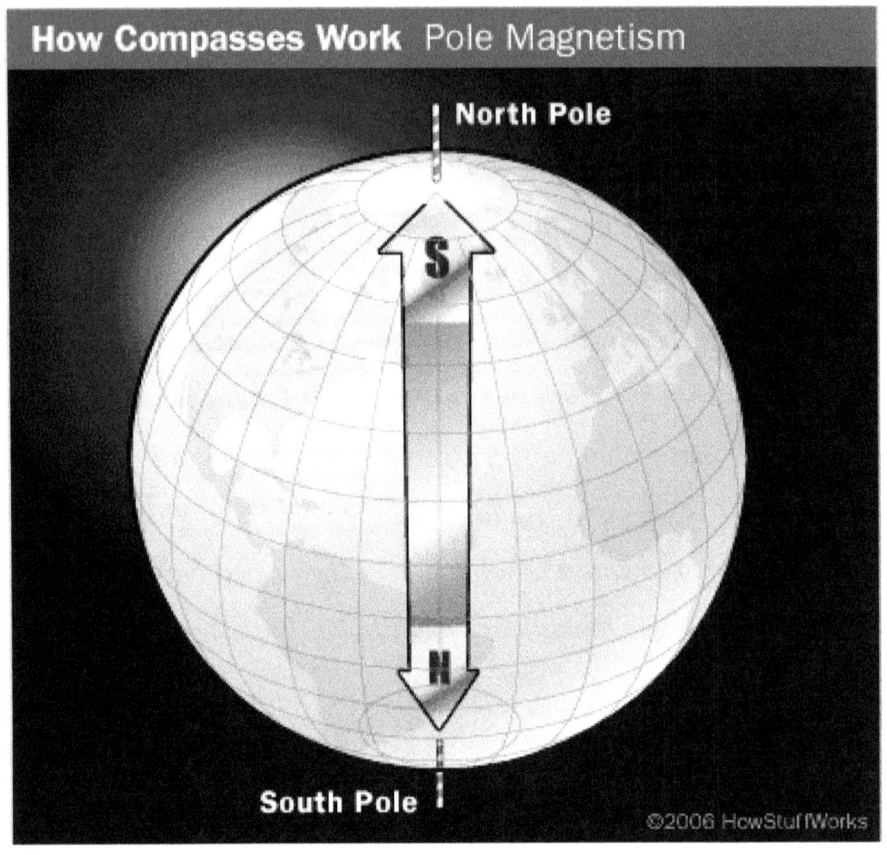

Picture [53]

Let me remind you again that it is because of the Earth's magnetism that we are able to find our way around the globe with the use of a compass. The animal kingdom relies on a built-in compass to navigate across the planet. Here again we can see how crucial the earth's magnetic field is for all life on Earth. More and more often in recent times, a question has been asked throughout the scientific world: What if the Earth's magnetic field reverses its polarity? Today it is pointing north; but what if it suddenly or in the near future starts pointing south?

Would a dramatic change in the Earth's magnetic field affect creatures that rely on it during migration? An article by Peter Tyson, editor in chief of Nova online, enlightens us:

"When I learned recently that our planet's magnetic shield is rapidly weakening and may be ready to reverse its polarity, causing compass to point south, I immediately wondered what that would mean for leatherbacks and the many other species that use the magnetic field to orient themselves and find their way around. Could they withstand a significant dwindling of the field's strength or even a reversal? Or might extinctions, perhaps mass extinctions, be in the offing?"[54]

Okay, let's put aside any consideration that the animals would be able to adjust to weakness or a reversal of the Earth's magnetic field. Let's just discuss the effect of a pole shift on us as humans. Based on the scientific discovery of the use of the pyramids, the intensity of the Earth's magnetic field is linked to our climate, resulting in climate change. To give you an idea of what may start to happen, here is an excerpt from an article entitled "Heat waves, floods and storms" by msnbc.com staff on this subject.

"Top international climate scientists and disaster experts meeting in Africa have a sharp message for the world's political leaders: Get ready for more dangerous and unpredictable weather caused by global warming... This is the first time the group of scientists has focused on the dangers of extreme weather events such as heat waves, floods, droughts and storms..."[55]

Wisdom should always compel us to worry when a natural disaster strikes our neighbors, and to look for its causes and learn how to prevent it, because that same disaster may someday happen to us.

A 2011 article entitled "Climate Change migration warning issued through report" by Pallab Ghosh, Science correspondent for BBC News reports:

"The government-commissioned report warns of potential humanitarian disasters because of climate change... The government's chief scientist, Professor Sir John Beddington, who commissioned the study, said that environmental change would hit

the world's poorest the hardest and that millions of them would inadvertently migrate toward, rather than away from, areas that are most vulnerable... [These people] will be trapped in dangerous conditions and unable to be moved to safety."[56]

Another BBC News Science & Environment article, entitled "IPCC: Climate impact risk set to increase: The risk of extreme weather is likely to increase if the world continues to warm, say scientists." It reports:

"There have been statistically significant trends in the number of heavy precipitation events in some regions... Average tropical cyclone maximum wind speed is likely to increase, although increases may not occur in all oceans basins...that small island-as well as mountainous and coastal-settlements were likely to be particularly vulnerable as a result of sea-level rise and higher temperatures, in both developed and developing nations... A study published in 2009 showed that hurricanes in the North Atlantic were more frequent than in the previous 1,000 years, and while the authors said the current level of activity was unusual, they stopped short of suggesting there was a direct link with a warming world..."[57]

Floods

Heat Waves

Professor Mike Hulme, professor of climate change at the University of East Anglia, UK, states that a warming world would create a greater risk of extreme weather, but it would be difficult to pinpoint what events were the results of greenhouse gas emissions. Hulme also states that impacts of climate extremes have a cumulative effect, which has not been properly assessed.

The IPCC, the leading international body for the scientific assessment of climate change was established by the UN Environment Program and the World Meteorological Organization (WMO) in 1988. Its main task has been to produce regular Assessment Reports.

So, the questions are: Is it too late and if not, what can we do?

Would you wish to play a role in saving the planet, guaranteeing a future to future generations?

Humbly, I submit to your attention my personal quest, my deep preoccupation for the future stretched by the premonitory dreams that have been given to me (the scientific goals of the pyramids!)

We will probably soon be reading more dramatic news

reports relating to climate change. The dream-inspired revelations given to me regarding the scientific purpose of the pyramids are in your hands. Perhaps you will decide that it is up to you to play a role in saving the planet.

I call on the United Nations to come together as one, to work together hand in hand, regardless of political or social differences, to learn, deepen and master the scientific power (a natural source of energy) behind the pyramids. Learning to harvest this new type of natural energy from the sun using the now discovered high technology of the pyramids will eventually change the way we perceive energy.

It is our duty to think of future generations. If we don't act now, their problems will be much more dramatic and catastrophic.

NOTES OF WISDOM

My revelations of the scientific purposes of the pyramids clearly point to solutions to the problem of climate change. It may take the world a few months or years to fully comprehend, to realize the a, b, to z of the construction and operation of the pyramids, and put these insights to good use.

Perhaps Planet Earth and the myriad life forms that live here will be grateful for the knowledge of the pyramids, which we now understand to be a "battery," and it will be a much better place even in this cycle of Kali Yuga (Iron Age).

Personally, I wrote this book in a short time, because I was compelled by the God Force, who gave me this assignment to bring it to your attention as fast as I could.

Although I have not released certain information given to me in the dream state by the God Force, as the world shows its commitment to learning and building pyramids, I will reveal the complete information.

In this booklet, finally, I only transmit the basic knowledge concerning our topic; there would be some additions to it in the future. I wish to inform the political and scientific world of my humble reason: to learn about the purpose of the pyramids.

I would also encourage you to teach its knowledge to our children, to ensure that their generation and the children of future generations continue to know about it and to master this knowledge. Given the discoveries and revelations I have witnessed, I wish to encourage its teaching.

And in the event that scientists or researchers would be designated to do this work, for example with government's sponsorship, I am ready to provide further information to the interested persons.

I share with millions of people the hope that the Earth will continue to be a beautiful training ground for souls as spiritual beings in transit, learning to experience the love of the Divine in each of us, to Love the Creator, the Great Spirit our Maker, who is the intelligence behind the making of all science, the Sun, the Earth and of the entire galaxy.

I am closing this scientific revelation of the pyramids by

sharing these three analogies with you.

One is a paragraph from the spiritual book entitled "In The Company of ECK Masters," by Phil Morimitsu. It's an out-of-body experience of Phil's with the spiritual guide called the Mahanta, where he went back in time during a crucial period in Atlantis. Our civilization knew the existence of the mythical lost civilization of Atlantis because of Plato's writings, notably in his book, "Timaeus and Critia."

At the time of this ultimate spiritual experience, Phil met a philosopher and wise man who told him:

"You see, every organism in this physical universe has an aura about it that acts as protection on the outer as well as on the Inner planes. Each man has an aura; even the planet has one. The health of this aura depends on a balanced inflow and outflow of the ECK current [the God Force, the Holy Spirit, the Sound Current, etc.]. *Inflow is needed to sustain life and feed new energy into it. The outflow is needed to give back to the outer world, so the process can repeat itself. Just as you and I must inhale and exhale air to survive.*

"If you inhale too much, without exhaling the commensurate amount, you will become bloated. The air must come out in one

manner or another, or you die. Likewise, if you exhale too much without inhaling properly, you will not gain enough air to survive, and death must also occur." [58]

The second is written by me to give you an analogy to understand the Earth and its magnetic field.

When a woman is pregnant, she feeds her baby from the food she eats and the water she drinks. If she stops eating and drinking, the baby will survive by eating the reserves inside the mother's body. And if the mother persists in not feeding herself, the baby will eventually die, which can also cause the death of the mother since her body becomes weaker as well without adequate nourishment.

As we take from the Earth, so must we also learn to give back to it in order to keep it alive and livable.

The third and last one is a profound reflection by Ignatius Donnelly, as shared here in his biography prepared by Lisa Dudek, Spring 2006:

"He was born-November 3, 1831, in Philadelphia, Pennsylvania. Ignatius became one of the best known and most forgotten figures in the political reformist age of the 1800s. He

was involved in the leadership or party policies of all of the major independent political parties active in the Midwest from 1870 until his death in 1891. In additional to having a busy political career, Donnelly was a prolific public speaker, the founder and editor of several newspapers, and the author of several books and novels. His life was spent pursuing many careers, often simultaneously. Furthermore, each enterprise Donnelly undertook was afforded the same vigor and tenaciousness as all of the others. [59]

In his book entitled, "Atlantis, the Antediluvian World," he stated:

"In religion the Atlanteans had reached all the great thoughts which underlie our modern creeds. They had attained to the conception of one universal, omnipotent, great First Cause. We find the worship of this One God in Peru and in early Egypt. They looked upon the sun as the mighty emblem, type, and instrumentality of this One God. Such a conception could only have come with civilization. It is not until these later days that science has realized the utter dependence of all earthly life upon the sun's rays:

"All applications of animal power may be regarded as

derived directly or indirectly from the static chemical power of the vegetable substance by which the various organisms and their capabilities are sustained; and this power, in turn, from the kinetic action of the sun's rays.

"Winds and ocean currents, hailstorms and rain, sliding glaciers, flowing rivers, and falling cascades are the direct offspring of solar heat. All our machinery, therefore, whether driven by the windmill or the water-wheel, by horse-power or by steam—all the results of electrical and electro-magnetic changes-- our telegraphs, our clocks, and our watches, all are wound up primarily by the sun.

"The sun is the great source of energy in almost all terrestrial phenomena. From the meteorological to the geographical, from the geological to the biological, in the expenditure and conversion

of molecular movements, derived from the sun's rays, must be sought the motive power of all this infinitely varied phantasmagoria." [60]

To read between the lines of this book, you may understand that **the way to save the planet is to avoid the extinction of its energizing Earth Core**.

However the civilization of Atlantis had achieved it by keeping pyramids active. Others who had probably also achieved it: the Pharaohs of Egypt, the kings of Sudan, the peoples of the Central America and the North, and those of Europe and elsewhere that had access to this secret wisdom.

Are you still skeptical? Allow me to ask you: why would the Pharaohs and the kings of Sudan permit the construction of 350 pyramids in their neighborhood if they were only simple burial vaults? If you observe the question from a scientific viewpoint instead of a religious perspective, you will be able to comprehend that the Pharaohs were protecting or helping their territory, Egypt, against something dangerous.

In Chapter 10 of the book "The Complete Pyramid Sourcebook" of John DeSalvo, Ph.D, one finds this observation:

"The research conducted in these large fiberglass Pyramids was coordinated and carried out by the following institutions in Russia and the Ukraine...Using radar, the "Scientific and Technological Institute of Transcription, Translation and Replication" in Kharkiv, Ukraine confirmed what they called an ionic formation up to 2000 meters above the Pyramid and a width of 500 meters...Statistics have shown that seismic activity diminishes in areas where Pyramids are built. It has been shown that instead of one powerful Earthquake occurring, hundreds of tiny ones occur instead."

It is one of the fundamental pieces of evidence in my book. Of course, this scientific conclusion by Russians researchers and Ukrainians only confirmed an aspect of the low tension of the pyramids that I had indicated.

Other scientific research seems to sustain the idea, according to which the cosmic energy converted through towers of stones, constructed from rocks and granite as well as mica-schist, and other natural elements, must have played an important role in the improvement of agricultural production techniques and therefore on farming, but more precisely on the fertility of soils.

The pyramid builders also had developed knowledge

regarding the influence of cosmic energy on vegetation. These scientific discoveries are mentioned in the works of Dr Philip Callahan, entomologist, radio engineer and author of many books and scientific articles, as related by the writer and researcher, Edward F. Malkowski, in his book entitled: *Ancient Egypt 39,000 ECB - the History, Technology, Philosophy and of Civilization X,* In the chapter entitled "Stone Towers as Energy Conductors," a book that I will recommend to any person interested in furthering or deepening their knowledge on this aspect of the science of cosmic energy on plants.

In the same context, in the chapter titled: " Fertilization with the help of the Pierre -- Fertilizing-"Stone with," the work done by the inventor and physicist John Burke, demonstrates that the use of the low shape of energy (the ions) in agriculture could improve the fertility of plants: plants grow more quickly and are more productive without having to resort to chemical fertilizers, etc. Many universities and companies that produce agricultural seeds collaborate with this process of "Molecular Impulse Response." Would it give a scientific explanation as to the presence of megalithic sites on our planet?

Would the science of magnetoculture and electroculture,

which was first proposed to the scientific community by l'Abbé Bertholon of France, be derived from the primitive 'model' of the pyramids?

Is there any other scientific connection between the science of the pyramids, the Nile River and the economic (agricultural, social, etc) prosperity of ancient Egypt?!

You are free to reach your own conclusions on the results of my research. But there is one point that you cannot refute: the presence of *Mica* in the Pyramid of the Sun. This mineral by its present use proves that the pyramids were tools for non-polluting energy production and renewable energy.

The "model" of the pyramids is a basis for a natural scientific solution in order to solve a lot of our energy problems on Earth, and in particular to prepare us to face the effects of climate change.

Let's recall the words of this famous philosopher:

"When it is not in our power to determine what is true, we ought to follow what is most probable."
— René Descartes

Switzerland, August 31, 2015 1:35 A.M

To the world leaders and the International Scientific Community, etc.

Subject: The Science of the Pyramids – The South Atlantic Anomaly (SAA) and Global Atmospheric Variations (Climate Change)

Open Letter

Dear Sirs and Mesdames:

Humanity can only survive the escalating atmospheric variations (climate change) if all countries combine their efforts to understand the real causes behind these variations. This is a moral and ethical obligation: it is our job to ensure the survival of the environment and protect the existence of future generations, as our ancestors did for us.

As founder of a non-profit organization, my investigations with the Pyramids prove that they are scientific instruments that were used in ancient times to resolve the same natural atmospheric variations that we are dealing with today that are leading to climate change. In 2011 in Washington State, through my foundation, Classical Music for Children, I transcribed the basics principles of these discoveries in my second book, *The Pyramids' Mysteries Resolved: Scientific Solutions to Challenges Regarding the Earth's Magnetic Field and Climate Change.* I informed President Barack Obama of the cause and solutions in 2011, followed by other world leaders in 2012, but my efforts have been in vain. Carbon dioxide buildup remains the primary agenda explaining the temperature anomalies – yet that is not the real cause.

Unable to get the attention of government leaders, and due to the strategic importance of the issue of climate change, I extended my call for help to European, Asian and African leaders, and was briefly able to share my findings with two European and one South American President in recent months. Today marks a new call to each of you, because climate change and the need for the information that my foundation is disclosing to humanity is more acute than ever.

The scientific community has established knowledge that atmospheric variations leading to climate change are caused by the weakening of the earth's magnetic field, rather than the increase of carbon dioxide (CO_2). However, I also point out that the process is accelerated by what is already known as "The South Atlantic Anomaly" (SAA), which could explain the extreme temperatures, both warm and cold, in various parts of the planet.

What is the South Atlantic Anomaly? NASA says:

"The Earth is surrounded by a pair of concentric donut-shaped clouds called the Van Allen radiation belts which, like magnetic bottle, store and trap charged particles from the solar wind. They are aligned with the magnetic axis of the Earth, which is tilted by 11 degrees from the rotation axis of the Earth, and are not symmetrically placed with respect to the Earth's surface. Although the inner surface is 1200 - 1300 kilometers from the Earth's surface on one side of the Earth, on the other they dip down to 200 - 800 kilometers. Above South America, about 200 - 300 kilometers off the coast of Brazil, and extending over much of South America, the nearby portion of the Van Allen Belt forms what is called the South Atlantic Anomaly. Satellites and other spacecraft passing through this region of space actually enter the Van Allen radiation belt and are bombarded by protons exceeding energies of 10 million electron volts at a rate of 3000 'hits' per square centimeter per second. This can produce 'glitches' in astronomical data, problems with the operation of on-board electronic systems, and premature aging of computer, detector and other spacecraft components.

"The Hubble Space Telescope passes through the 'SAA' for 10 successive orbits each day, and spends nearly 15 percent of its time in this hostile region. Astronauts are also affected by this region which is said to be the cause of peculiar 'shooting stars' seen in the visual field of astronauts."
- From NASA web site:
(https://heasarc.gsfc.nasa.gov/docs/rosat/gallery/misc_saad.html)

I have compiled a 16-minute video titled "Pyramids' Sciences and Climate Change 4th Anniversary and the South Atlantic Anomaly (SAA)", to give you a clear understanding of the South Atlantic Anomaly's effect on the current atmospheric variations or climate change.

My argument is that the way climate change is approached today is incorrect. The presence of the pyramids all over the planet is a key to the great, lost ecological science used and left by ancient civilizations who possessed a working knowledge of atmospheric variations and their recurrence. Thus, to guarantee the survival of the planet. I implore the scientific community to look into the solution the pyramids offer for protection of the environment.

It is now our responsibility to learn about this science of the stone, which is partially detailed in my book.

As I urged in my open letter to world leaders in 2012, a consortium of international scientists must be put into place to further investigate, with my help, the pyramids as science.

I hope this information will be received with wisdom by you and other trusted leaders and governments. My only objective since 2011 has been to enlist greater consideration and attention to the environmental solution offered by the ancients that we may use again today for the love and security of the future of humanity.

Sincerely,

Christian Bernard Magnongui
Founder, Independent Researcher at
www.classicalmusicforchildren.org
(The Pyramid project) and Climate Change
I administer a humanitarian blog page at:
http://info80418.wix.com/bernardmagnongui

Other articles related to this file:

1- Dr. Judith Curry's Testimony before the House of Representatives Committee on Science, Space and Technology hearing on the President's UN Climate Pledge.
2- Another Look at the Meaning of the 4th of July 2015 – The Sciences of the Pyramids and Climate Change
3- Letter to the Department of Homeland Security Regarding the Science of the Pyramids and Climate Change for the USA

Etc.

ABOUT THE AUTHOR

A native of the Republic of Congo, Brazzaville, Bernard Christian Magnongui studied economic and computer science until his plans were interrupted by civil wars. He then served as assistant to the president of the Karate Association of South Africa in Johannesburg before moving to the United States in 1999.

He furthered his studies in computer technology in 2000 and worked as a loan officer and real estate assistant until the mortgage industry collapse in 2008. Then, guided by a Dream, he became a direct care specialist for the mentally retarded in Austin, Texas. Looking for ways to lighten up their spirit, he received his first wake-up call, to use classical music for its healing powers.

Later, as a volunteer math tutor for youth in Opelousas, Louisiana, a second wake-up call delivered the inspiration to develop a classical music program to help students. It was a resounding success, in large part, Bernard realized, because of the support of parents and educators who believed in the importance of music or witnessed its effect upon their children.

A later Dream, in November 2009, detailed the research

needed to write and publish his first book—*The Power of Musical Sound: how music affects our state of mind, health and society*—as a tool to encourage such programs.

Bernard followed up on his inspiration by founding the Classical Music for Children Foundation, a non-profit organization dedicated to promoting the positive effects of sound and music in educational environments.

In August 2011, one month after moving from Boston, Massachusetts to Washington State, Bernard had an out of body experience in the Giza Pyramid in Egypt and a series of Dreams revealing the scientific purpose of the Pyramids.

With a spirit of adventure, after almost seven months without work, he chose to dedicate his time to investigating the subject until he was able to confirm his intuitive guidance regarding the scientific purpose of the Pyramids and how they are linked to climate change.

A MESSAGE FROM THE FOUNDATION

As of today December 01, 2012, none of the above mentioned news organizations, foundations and political leaders have contacted me to discuss or to learn more about this project despite my various attempts to contact them. And because of this silence especially from the USA where I live, and seeing the increasing challenges of climate changes, I have decided to travel to Europe and to Africa to seek support and sponsors in order to materialize this Dream for the benefit of our children and grandchildren; for the benefit of planet Earth.

The scientific revelations of the Pyramids I brought forth also come with the key of the Pyramid. Once the conditions are fulfilled by some chosen world leaders and international scientists, I will share the second gift to the world, a gift that will prove the validity of my scientific revelations of the Pyramids. To Contact me for a quick response, email us at:

www.ClassicalMusicForChildren.org (The Pyramid Project)

Seattle, February 03, 2012

PRESS RELEASE FOR IMMEDIATE RELEASE

The Classical Music for Children Foundation, a non-profit organization, announces the release of a new book, "THE PYRAMIDS' MYSTERIES RESOLVED: SCIENTIFIC SOLUTIONS TO EARTH'S MAGNETIC FIELD AND CLIMATE CHANGE," written by Christian Bernard Magnongui.

What follows is the text of a letter sent by the author to various world leaders outlining the author's call to action. Copies of this letter have been sent, along with a copy of Mr. Magnongui's book, by mail and electronically to 23 Presidents, Chancellors and Prime Ministers via their official channels and embassies (see list below):

Open Letter

Dear Presidents, Chancellors and Prime Ministers,

Thank you for receiving my newly published book entitled, The Pyramids' Mysteries Resolved: Scientific Solutions to Earth's Magnetic Field and Climate Change. The subject of climate change covered in the book is an urgent matter, and demands your immediate attention and consideration.

The information in this book is of critical importance to the wellbeing of the earth at this time. When you read this book you will understand on a deeper level what is actually happening to our climate, and you will discover the key to technology for resolving serious climate change issues we are currently facing and will face in the near and far future.

To date, I have published this book and have sent free copies out to some key people (Newspaper editors, CEOs, Professors, etc) using funds lent to me by friends and from sales of my two books. It is urgent that I keep getting copies of this book into the hands of people around the world who can help with the climate change discoveries contained in it.

My goal is also to travel with a team to Egypt and other countries where pyramids continue to deepen our understanding of the connection between pyramids and climate change, and to help scientists find the missing information they need to resolve this critical issue.

To do this work I see that two things must be done:

1. Establish an international consortium of scientists to assist in the work.
2. Obtain funds so that my team and I can continue to send copies of the book to key people, including traveling to pyramids to learn more and beginning to study the process of incorporating the pyramids' scientific purpose.

To date, I alone have donated my time and financial resources to bring these discoveries to the public, but that is not enough. I know that much more needs to be done to bring my research to the right people and bring them together as a united team for the purpose of helping resolve climate change issues through the scientific operation of the pyramids. I need your help in this important cause.

If you decide to support this important work, you will have a voice in how the resources are to be used. An international consortium or oversight committee will be formed for this purpose. The importance of spreading this information dictates the need to go worldwide to bring scientists together to work on this project.

Currently I have an established foundation called the "Classical Music for Children Foundation." My plan is to make the Pyramid Project a branch of that foundation for three reasons:

1.   To allow immediate action versus taking time and money to establish a new foundation.
2.   To save resources as a subsidiary of the Classical Music For Children Foundation.
3.   To enhance the future of the children of the world to resolve the climate issue as soon as possible.

It is critically important to establish a consortium of international scientists and receive sponsorships and funds in order to continue this mission. Donations, as well as suggestions and support can be sent to:

The Pyramids: Scientific Solutions to Earth's Magnetic Field & Climate Change.
PO BOX: 1240
Seattle, Washington 98111-1240
U.S.A
Email: Info@ClassicalMusicForChildren.org
www.ClassicalMusicForChildren.Org

I can be contacted by email, phone, or mail with invitations by serious and responsible democratically-elected governments, individuals, businesses or organizations to discuss this request.
To those who participate with donations of time, research and

money, an accounting of all funds received will be made on a regular basis, and may be tax deductible.

In addition, the consortium will also need office space, equipment, and technological expertise. Everyone who contributes will be kept informed of pertinent findings from the research.

I appreciate your understanding and I believe that once this project is given the consideration it deserves and has started to produce results, it will be a blessing to our earth and to future generations of our children. That is the goal of The Pyramids' Mysteries Resolved: Scientific Solutions to Earth's Magnetic Field and Climate Change.

(END OF LETTER)

A complimentary copy of Mr. Magnongui's book is enclosed for your review. Mr. Magnongui can be contacted as follows:
Christian Bernard Magnongui
Email: Info@ClassicalMusicForChildren.org

PO BOX: 1240,
Seattle, Washington 98111-1240
U.S.A
www.ClassicalMusicForChildren.org.

1. United State of America: President Barack H. Obama
2. Canada: Prime Minister Stephen Harper
3. Kingdom: Prime Minister David Cameron
4. France: President Nicolas Sarkozy
5. Japan: Prime Minister Yoshihiko Noda
6. Mexico: President Felipe de Jesús Calderón Hinojosa

7. Korea: President Lee Myung-bak
8. President Ollanta Moisés Humala Tasso
9. Chancellor Angela Dorothea Merkel
10. Africa: President Jacob Zuma
11. Acting President Mohamed Hussein Tantawi
12. President Pratibha Patil
13. President *Dilma Rousseff*
14. President Micheline Anne-Marie Calmy-Rey
15. Ghana: President Professor John Evans Atta Mills
16. President Giorgio Napolitan
17. Republic of Congo: President Joseph Kabila
18. President Abdoulaye Wade
19. Prime Minister Helle Thorning-Schmidt
20. Prime Minister Julia Gillard
21. Prime Minister Elio Di Rupo
22. Sweden: Prime Minister Fredrik Reinfeldt
23. Nigeria: President Goodluck Jonathan

CC: United Nations General Secretary; CNN News; FOX News; ABC News; BBC News; CBS News; Boston Globe; Star Tribune; AFP News; Reuters News; Coasttocostam.com; NPR; The Seattle Times; Boston Globe; NYTimes; Wall Street Journal; Chicago Tribune; Houston Chronicles; Los Angeles Times; NAACP; Native American Associations; Asian Associations; NASA; European Space Agency; Environmental Defense Fund; World Meteorological Organization; IPCC Secretariat; European Union; African Union Secretary; National Geography; As well as other national news organizations of the countries listed above.

(The translation of this letter in French, German and Spanish will be made available to download for news only at www.ClassicalMusicForChildren.org)

REGARDING MUSIC EDUCATION FOR OUR CHILDREN

Please help spread the knowledge and join my cause to provide youth with a sound classical music education. By enjoying and sharing our book entitled "**The Power of Musical Sound: How Music Affects Our State of Mind, Health and Society,**" sold at Amazon.com, Barnes & Noble as well as through our non-profit website.

In purchasing these books, you will help the foundation to raise funds for our grant program to foster classical music training for children in underprivileged communities in the U.S.A and around the world.

To achieve our goal, help us sell 1 million copies of our book "The Power of Musical Sound."

We offer a discount for educational or non-profit organizations that purchase large quantities of this book, or to anyone willing to donate quantities of this book to those organizations. Both paper and electronic versions of this book are available on our website.

We are seeking assistance for translations as well. We intend for the book to be translated into French, Spanish, German and Arabic.

Thank you for your interest, care, and willingness to share the spirit of music in your heart!

– Bernard Christian Magnongui, Founder

www.ClassicalMusicForChildren.org

CLASSICAL MUSIC FOR CHILDREN FOUNDATION

Non-profit Organization Dedicated to Classical Music Education for Underprivileged Children

MISSION STATEMENT

The purpose of the Classical Music for Children Foundation is to provide education for parents and students about the effects that music has on health and well-being. This includes not only the sounds of music but also the words of music, and the ways they impact creativity and imaginative faculties.

Our goal is to provide training and classes for children in how to read, write, and compose classical music. We plan to target underprivileged children, but we hope to focus on all children in these offerings.

In this process, we hope to raise public awareness about the significance of classical music education as a primary instrument to raise the quality of education in the USA.

Our plan is to expand our activities around the world.

Note: No substantial part of the foundation's activities involves attempts to influence legislation. The Foundation does not intervene in political campaigns.

This foundation is supported by the income from the sales of our books including the entitled, "The Power of Musical Sound: how music affects our state of mind, health, and society."

We are in the process of establishing a 501c3 status.

Classical Music for Children Foundation

www.classicalmusicforchildren.org
ClassicalMusicForChildren@Gmail.com

CREDITS & WORKS CITED

[1] Galileo Galilei http://www.crystalinks.com/galileo.html

[2] http://www.goodreads.com/quotes/542062-the-world-will-not-be-destroyed-by-those-who-do

[3] Today In Science History http://todayinsci.com/home.htm

[4] Benjamin Banneker
http://www.bnl.gov/bera/activities/globe/banneker.htm

[5] Letter to Dr Priestley, 8 Feb 1780. In *Memoirs of Benjamin Franklin* (1845), Vol. 2, 152.

http://todayinsci.com/F/Franklin_Benjamin/FranklinBenjamin-Quotations.htm

[6] http://www.the-philosophy.com/montesquieu-quotes

Montesquieu, Baron. The Spirit of the Laws. Electronic Text Center, University of Virginia Library
http://etext.lib.virginia.edu/toc/modeng/public/MonLaws.html

[7] Vrooman, Jack Rochford. Rene Descartes: A Biography. New York: G.P. Putnam's Sons, 1970. (Page 56, 58)

[8] La philosophie de Descartes

http://la-philosophie.com/philosophie-descartes

Goodreads
http://www.goodreads.com/author/quotes/36556.Ren_Descartes

[9] The Tuthmosis IV Dream Stele"

http://ib205.tripod.com/sphinx_dreamDream.html

[10] Rational Nation USA
http://rationalnationusa.blogspot.fr/2012/03/montesquieu.html

[11] The Mahanta, the Living ECK Master
http://www.eckankar.org/Harold/index.html

[12] Egyptians Pyramids

http://www.egyptian-PyramidPyramids.co.uk/the-great-PyramidPyramid.htm

[13] Mystery of the Sun (NASA)
http://www.nasa.gov/mission_pages/sunearth/news/mystery-sun.html

[13A] Le Soleil- Wikipédia http://fr.wikipedia.org/wiki/Soleil

[13B] Magazine Science http://www.paperblog.fr/2699718/2-699-999-990-000-decimales/

[14] Dimensions and Mathematics of the Great Pyramid
http://www.theglobaleducationproject.org/egypt/studyguide/gpmath.php

[15] David Linden: *Handbook of Batteries. Second Edition*, 1995. Page 1.3

[16] Car and deep cycle battery http://www.batteryfaq.org/

[17] Christopher, Dunn. *"The Evidence Leading up to Gantenbrink's 'door'"*. Web. September 15, 2002.
http://www.gizapower.com/ShaftEvidence.htm

Christopher, Dunn. *"Evidence of Ancient Electrical Devises found in the Great Pyramid?*" Web. June, 2 2011.

http://gizapower.com/Anotherrobot.htm

[18] The King Chamber Diagram
http://www.timstouse.com/EarthHistory/Egypt/GreatPyramidPyramid/diagrams.htm

[19] The Queen Chamber Diagram
http://www.timstouse.com/EarthHistory/Egypt/GreatPyramidPyramid/diagrams.htm

[20] Battery Council International. "How a Battery is Made".
September 2010.
http://www.batterycouncil.org/leadacidbatteries/howabatteryismade/tabid/107/default.aspx

[21A] Histoire d'un mystère : l'intérieur de la Terre – Planète - Terre'- Vincent Deparis- Maison des Sciences de l'Homme - Alpes, Grenoble.

http://planet-terre.ens-lyon.fr/planetterre/XML/db/planetterre/metadata/LOM-modeles-interieur-terre.xml

[21] The Earth Core by Geologist
http://journalgeology.com/v1/structure-and-composition-of-earth/

[21B] Couches Géologiques de la terre Wikipédia.

http://fr.wikipedia.org/wiki/Terre

[22] Dr. Ken Rubin, Assistant Professor at the Department of Geology and Geophysics, at the University of Hawaii

http://www.soest.hawaii.edu/GG/ASK/earths_core.html

[22A] Observation mathématique de la pyramide de Khéops

Wikipédia Jean-Philippe Lauer, Le mystère des pyramides, 1988, p.234.

http://fr.academic.ru/dic.nsf/frwiki/1253394

[23] [24] Dimensions and Mathematics of the Great Pyramid
http://www.theglobaleducationproject.org/egypt/studyguide/gpmath.
php

[24A] God's Time Capsule

http://godstimecapsule.com/chapter-2/

[25] Great Pyramid of Giza Research Association.
http://www.gizaPyramidPyramid.com/gip2.htm

[26] US geological survey. www.mindat.org

[27] Pierre and Marie Curie in the laboratory
http://en.wikipedia.org/wiki/File:Pierre_and_Marie_Curie.jpg

[27A] Frank Dorland. Holy Ice. Galde Press, Inc. St Paul,
Minnesota, 1992. Page 119-120

[27B] http://www.samuellaboy.com/English/prologue.htm

[28] Frank Dorland. Holy Ice. Galde Press, Inc. St Paul, Minnesota,
1992. Page 119-120

[28] http://www.gizapower.com/ShaftEvidence.htm)

[29] John Meurig Thomas, FRS: Michael Faraday And the Royal
Institution: The Genius of man and Place. IOP Publishing Ltd, 1991.
Page 41

[30] Smithsonian Science – Michael Faraday

http://smithsonianscience.org/2012/06/michael-faraday-1791-1867-chemist-and-physicist-at-the-royal-institution-of-great-britain/

[31] Today in Science History
http://www.todayinsci.com/F/Faraday_Michael/FaradayMichael-Quotations.htm

[31A] Une nécropole de 35 pyramides découverte au Soudan – Futura Science

http://www.futura-sciences.com/fr/news/t/paleontologie/d/une-necropole-de-35-pyramides-decouverte-au-soudan_44508/

[31B] Les communautés originaires du Soudan.

http://www.darnna.com/phorum/read.php?12,147187,page=2

[31C] Le Mica - Dictionnaire Minéralogique

http://fr.wikipedia.org/wiki/Mica

[31D] Pyramides du Soleil Ses Mathématiques Les pyramides de la chine.

http://www.ancient-wisdom.co.uk/pyramids.htm

[31E] La Pyramide de Cestius (Italie)

http://www.thesportscoupe.com/place3/index.php?lang=fr&zr=20&zc=IT&za=Latium&zs=Rome&zl=Rome&mode=5&m=6&pl_id=201203230025

[31F] Pyramides en Bosnie

http://secretebase.free.fr/civilisations/ruines/bosnia/bosnia.htm

[32] Journal of Serendipitous http://www.jsur.org/history/Oersted

[33] Today in Science
http://todayinsci.com/O/Oersted_Hans/OerstedHans-Quotations.htm

[34] Gillian Turner. *North Pole, South Pole: The Epic Quest to Solve the Great Mystery of Earth's Magnetism.* New York: The Experiment, LLC, 2011. Page 79-80

[35] Rick Groleau. "When Our Magnetic Field Flips". Web. November 18, 2003.

 http://www.pbs.org/wgbh/nova/earth/when-our-magnetic-field-flips.html

[36] Magnetic Fields History http://www-spof.gsfc.nasa.gov/Education/whmfield.html

[37] Jin Marvin Herndon's Origin of Earth Magnetic Field.
http://www.nuclearplanet.com/Herndon's%20Geomagnetic%20field.html

[38] National Center for Urban and Industrial Center (US): *Manual of Septic Tank practice. 1967. Page 27*

[39]Inspectpedia
http://www.inspectapedia.com/water/Water_Pollution_15.htm

[40] Dig It Excavating Inc. "It's All Connected An overview of On-site Septic Systems – YouTube". Web.
http://www.youtube.com/watch?v=i6yFtzkV34Q

[41] http://www.a-1hotels.com/eg/history/html/PyramidPyramids_of_giza.html

[42] Gillian Turner. *North Pole, South Pole: The Epic Quest to Solve*

the Great Mystery of Earth's Magnetism. New York: The Experiment, LLC, 2011. Page 99

[43] Physics of Energy & the Environment- PHYS 161 Lecture 10

http://hendrix2.uoregon.edu/~dlivelyb/phys161/L10.html

Physics of Energy & the Environment- PHYS 161 Lecture 1

http://hendrix2.uoregon.edu/~dlivelyb/phys161/L1.html

[44] Earth's Magnetic Fields and How It Protects Life

http://magnet-therapy-24.blogspot.com/2011/11/earths-magnetic-fields-and-how-it.html

[44A] Le Champs Magnétique Terrestre –

http://aurores-polaires.e-monsite.com/pages/vent-solaire-
        magnetosphere-aurore-polaire/1.html

[44B] [46] Aimants et Magnétisme

http://www.magnetosynergie.com/Pages-Fr/Aimants/FR-Aimants-
        04.htm

[45] The Dynamic Earth
        http://www.mnh.si.edu/earth/text/4_1_5_0.html

[46] Gilliam Turner *North Pole, South Pole: The Epic Quest to Solve
        the Great Mystery of Earth's Magnetism* Page 113[47] Nola
        Taylor Redd. "Where'd Mars water go? May be
        underground". Technology & Science. Web. November 24,
        2011.

http://www.msnbc.msn.com/id/45139654/ns/technology_and_scienc
        e-space/t/whered-all-mars-water-go-maybe-
        underground/#.TuKrRrLTr3U

[48] Proceedings of National Academy of Sciences. "*NASA: DNA
        Found on Meteorites Indicates Life May Have Originated in
        Space*". Web. August 9, 2011.
        http://www.ibtimes.com/articles/195073/20110809/nasa-dna-
        meteorites-building-blocks-life-on-earth-from-space.htm

[49] Gilliam Turner North Pole, South Pole. Page 244[50] Lewis
Page. "The Register" environment reporting magazine. Web.
November 24, 2011.
http://www.theregister.co.uk/2011/11/24/earth_core_silicon_perhaps/

[51] Richard Black. "*Global warming 'confirmed' by independent*

study; The earth's surface really is getting warmer, a new analysis by a US scientific group set un in the wake of the "Climategate" affaire has concluded." Environment correspondent for BBC News. Web. October 20, 2011. http://www.bbc.co.uk/news/science-environment-15373071?print=true

[51A] The Revelation of the Pyramids
http://revelationofthepyramids.com/the-movie.php

[52] Marshall Brian. "*How compasses work*". How Stuff Works. Web. http://adventure.howstuffworks.com/outdoor-activities/hiking/compass1.htm

[53] How compass work?
http://adventure.howstuffworks.com/outdoor-activities/hiking/compass1.htm

[54] Peter Tyson. "*Would a dramatic change in the Earth's magnetic field affect creatures that rely on it during migration?*' Nova online. Web. November 18, 2003.
http://www.pbs.org/wgbh/nova/nature/magnetic-impact-on-animals.html

[55] Msnbc.com staff and news service reports. "*Heat waves, floods and storms: scientists warn world to prepare for extreme weather. Web. November 18,* 2011.
http://www.nbcnews.com/id/45353104/ns/us_news-environment/#.TuVOm7LTr3V

[56] Pallab Ghosh. "*Climate Change migration warning issued through report*". Science correspondent for BBC News. Web. October 19, 2011. http://www.bbc.co.uk/news/science-environment-15341651

[57] BBC News Science & Environment. "*IPCC: Climate Impact*

risk set to increase: The Risk of extreme weathers is likely to increase if the world continues to warm, say scientists." Web. November 18, 2011. http://www.bbc.co.uk/news/science-environment-15745408?print=true

[58] Morimitsu, Phil. "*In the Company of ECK Masters*". Minneapolis: Eckankar, 1987. Page 270

[59] Ignatius Donnelly – Online Biography prepared by Lisa Dudek, Spring 2006.

http://pabook.libraries.psu.edu/palitmap/bios/Donnelly__Ignatius.html

[60] Atlantis the Antediluvian World by Ignatius Donnelly, Page 277. http://www.gutenberg.org/dirs/etext03/ataw11h.htm#start

THE SCIENCES OF THE PYRAMIDS AND THE EARTH'S MAGNETIC FIELD

CLIMATE CHANGE

www.ingramcontent.com/pod-product-compliance
Lightning Source LLC
Chambersburg PA
CBHW051446170526
45166CB00001B/130